Special Publications 67

DENDROCLIMATIC STUDIES

Tree Growth and Climate Change in Northern Forests

Rosanne D'Arrigo
Nicole Davi
Gordon Jacoby
Rob Wilson
Greg Wiles

This Work is a co-publication between
the American Geophysical Union and John Wiley & Sons, Inc

AGU™
American Geophysical Union

WILEY

This Work is a co-publication between the American Geophysical Union
and John Wiley & Sons, Inc.

Published by John Wiley & Sons, Inc., Hoboken, New Jersey
Published simultaneously in Canada

Library of Congress Cataloging-in-Publication Data

D'Arrigo, Rosanne, 1954–
Dendroclimatic studies : tree growth and climate change in northern forests / Rosanne D'Arrigo [and four others].
pages cm
Includes bibliographical references and index.
ISBN 978-1-118-84872-2 (pbk. : alk. paper) 1. Old growth forests–Climatic factors–Northern Hemisphere. 2. Trees–Climatic factors–Northern Hemisphere. 3. Climate change mitigation. 4. Lamont-Doherty Earth Observatory. I. American Geophysical Union. II. Title. III. Title: Tree growth and climate change in northern forests.
SD387.O43D37 2014
577.3–dc23

2013050491

Cover image: Courtesy of Rob Wilson, Author, St Andrews Tree Ring Lab, University of St. Andrews, St. Andrews, UK. Sample 6000-7000 yrs old representing maximum pine woodland extent into the Cairngorms during the mid-Holocene optimum.
Cover design by Modern Alchemy LLC

Printed in Singapore

10 9 8 7 6 5 4 3 2 1

CONTENTS

PREFACE

As graduate students and research scientists over the past few decades, we have been fortunate enough to have the chance to travel to the far northern latitudes with each other and other colleagues from our laboratories and the wider tree-ring community. These forays to recover samples of old growth living and relict preserved wood for paleoclimatic studies involved searching for remote, undisturbed forests at treeline across northern North America and Eurasia. At these locations, the trees we sampled were at the limits of their survival and have served reliably as natural thermometers recording temperature conditions over the past several centuries or more. To visit such sites, for example in the far-flung Thelon River Sanctuary or the Coppermine River area of northern Canada, we would travel by charter plane from scheduled airports at such hubs as Yellowknife, Northwest Territories, already far from the beaten track. Our plane would then land on a river, dropping us with inflatable rafts and several weeks of camping gear, food, and full body gear for protection from the mosquitoes. We would then paddle downriver in our boats, stopping to sample any trees that caught our eye along the way. The possibility of a bear encounter was always in the back of our minds, and we saw our share. By day we searched for old trees, by night setting up camp and fishing for Arctic char and grayling for dinner.

We keep returning to some of these superb ancient sites to see how the trees have been faring over recent decades. Are the warmer temperatures inciting more growth? Or is warming causing drought stress and weakening the climate signal? These trees have much to tell us about the ever-changing response of the Earth's climate system to greenhouse warming and related environmental change, and so the forays will continue.

Rosanne D'Arrigo, *Tree-Ring Laboratory, Lamont-Doherty Earth Observatory, Palisades, New York, USA*

Nicole Davi, *Department of Environmental Science, William Paterson University, Wayne, New Jersey, USA*

Gordon Jacoby, *Tree-Ring Laboratory, Lamont-Doherty Earth Observatory, Palisades, New York, USA*

Rob Wilson, *School of Geography and Geosciences, University of Saint Andrews, St. Andrews, Scotland, UK*

Greg Wiles, *Department of Geology, The College of Wooster, Wooster, Ohio, USA*

ACKNOWLEDGMENTS

This project was funded by the National Science Foundation's Division of Environmental Biology's OPUS program grant DEB 09–48827 to Columbia University. TRL-LDEO northern tree-ring research was funded by the National Science Foundation Paleoclimatology and PaleoPerspectives on Climatic Change (P2C2) Programs and its previous initiatives, and NOAA's Climate Change and Data Detection Program and Paleoclimatology group, with additional support from NASA and the Department of Energy. We also thank the U.S. Parks Service, Parks Canada, the Canadian government, the State of Alaska, Bureau of Land Management, Inuit groups of Alaska and Canada, the U.S. Fish and Wildlife Service, the Arctic National Wildlife Refuge (ANWR), and many other programs and organizations that assisted with field work and logistics. We gratefully acknowledge contributors to the International Tree-Ring Data Bank (ITRDB, Glossary) for collecting and providing data, and Bruce Bauer of NOAA for online processing of the Northern Archive data. We thank the various reviewers for suggestions that improved this manuscript. Additional funding for R. Wilson was obtained from the European Union Millennium Project (Grant No. EU-017008-2-Millennium), for G. Wiles from the National Geographic Society and Keck Geology Consortium, and for N. Davi from the National Science Foundation postdoctoral fellowship program (contract AGS PRF 1137729). The American Geophysical Union (AGU) and Wiley publishers are gratefully acknowledged for publishing this monograph.

We also acknowledge various contributions to the northern archive by Kevin Anchukaitis, Laia Andreu-Hayles, Valerie Barber, David Barclay, Pieter Beck, Brendan Buckley, Chris Buckley, Oyunsanaa Byambasuren, Parker Calkin, Jobie Carlisle, Paolo Cherubini, Ed Cook, Ashley Curtis, Ts. Davaajamts, Jan Esper, Jim Feaver, Ben Foster, David Frank, Rosemary Free, Inez Fung, Scott Goetz, Moses Gostev, Heather Griffin, Ileana Ivanciu, Glenn Juday, Susan Kaplan, Osamu Kobayashi, Paul and Anna Krusic, David Lawrence, Nicolai Lovelius, Brian Luckman, Erika Mashig, Baatarbileg Nachin, Biligbaatar Nazad, Neil Pederson, John Sakulich, Fritz Schweingruber, Paul Sheppard, Oleg I. Shumilov, Olga Solomina, Bayambagerel Suren, Roy Thetford, Linda Ulan, Ricardo Villalba, Martin Wilmking, and David Yamaguchi.

Lamont-Doherty Earth Observatory Contribution No. 7753.

1

1. INTRODUCTION

1.1. Overview

This monograph provides a comprehensive synthesis of dendroclimatic research based on northern old growth sites in the forests of North America and Eurasia, as conducted by the scientists at the Lamont-Doherty Earth Observatory's Tree-Ring Laboratory (TRL-LDEO) of Columbia University over the past four decades (Figure 1.1).

We focus herein mainly on the latitudinal northern treeline (see Glossary), as well as other locations further south (such as the elevational treeline in Mongolia, and the Russian Far East) where trees are often sensitive indicators of past temperature variability. In keeping with the requirements of the "Opportunities for Promoting Understanding through Synthesis" (OPUS) program of the National Science Foundation that funded the writing of this monograph, we emphasize the research conducted by scientists at the TRL-LDEO. However, we place these studies in the context of the broader field of Northern Hemispheric dendroclimatology (Glossary), conducted by numerous other researchers and colleagues who have provided valuable and important insights into this topic over the years.

1.2. Basic Tree-Ring Principles

The research described herein adheres to the basic principles of dendrochronology, as outlined in introductory and general texts by Stokes and Smiley (1968), Fritts (1976), Schweingruber (1988), Cook and Kairiukstis (1990), Speer (2012), and others. The main premise of dendrochronology is the establishment of precise, high-resolution (annually-resolved) tree-ring chronologies, derived using the method known as cross-dating (references above; Glossary). The cross-dating technique is based upon the observation that there is a common climatic and environmental signal in the ring-width variations of samples of wood compiled from trees (of the same species) from the same site and

---

Dendroclimatic Studies: Tree Growth and Climate Change in Northern Forests,
Monograph Number 67. Rosanne D'Arrigo, Nicole Davi, Gordon Jacoby,
Rob Wilson, and Greg Wiles.

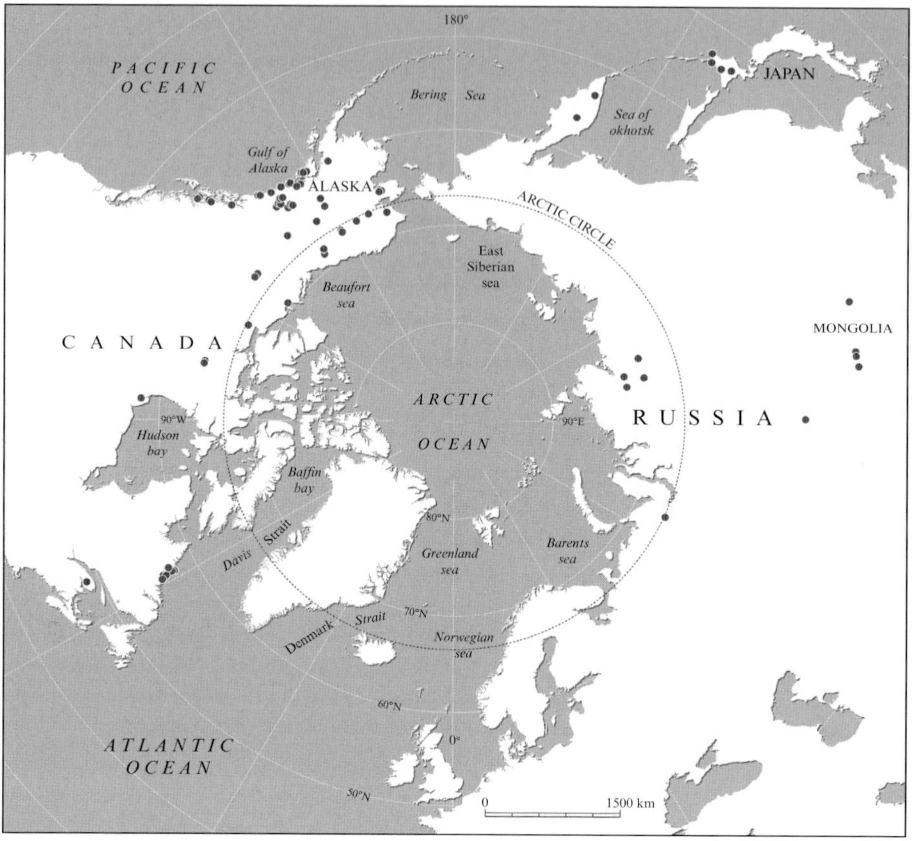

Figure 1.1 Locations of tree-ring chronologies sampled from northern forest sites over the past few decades by TRL-LDEO scientists and colleagues. Many more have been sampled by other scientists for these areas. For color detail, please see color plate section.

region. Relatively narrow (wide) rings are used to infer more adverse (favorable) environmental conditions for growth. When performed correctly, this method ensures that there are no dating errors resulting from anomalous growth patterns. These include, for example, false rings (Glossary); in which growth is slowed, resulting in thicker-walled cells, for a period during the growing season due to a particular adverse event, such as drought, or missing rings (in which radial growth is not laid down in a particular wood sample or tree due to an adverse event; Glossary). Although it has been suggested by Mann and colleagues (2012), based on tree-growth model simulations, that such missing rings can occur amongst *all* trees at a given site following major volcanic events (e.g., in 1258, 1815), there is no actual tree-ring evidence that this is in fact the case, as indicated in several subsequent presentations and publications (Anchukaitis et al., 2012a; Esper et al., 2013, in press; D'Arrigo et al., 2012a; D'Arrigo et al., 2001b; Journal of Geophysical Research, in press).

The science of dendroclimatology evolved from the need to understand past and present climate variability as well as the factors impacting tree growth and climate response on a range of spatial and temporal scales. Determination of how climate has varied in the past is also critically important for evaluating the sensitivity of the Earth's climate system to both natural and anthropogenic forcing. Yet, instrumental observations are limited in length and spatial coverage, particularly in many remote far northern regions, where station records may only span a few decades. Overcoming these limitations requires high-resolution, precisely-dated proxy data archives, such as tree rings, so that we may derive a long-term perspective for conditions during the recent anthropogenic era, during which profound and rapid changes are now taking place.

1.3. Polar Amplification of Global Warming and Impacts on Forests

The far northern latitudes are uniquely sensitive to climatic change, with warming in some regions at nearly twice the rate of many other areas of the planet (Figure 1.2) (Hansen et al., 2010; ACIA, 2005).

Alaska, for instance, has experienced a winter temperature increase of ~3.5 °C over the past ~50 years (NOAA Strategic Plan, 2010; National Academy of Sciences, 2011; Polar Frontiers Executive Summary, Grebmaier et al., 2011; Jansen et al., Intergovernmental Panel on Climatic Change [IPCC], 2007; U.S. Global Change Research Program—Regional Climate Impacts, Alaska Report, 2009; and AR5, www.ipcc.ch). Summer (Jun–Jul–Aug) temperatures in Arctic Alaska increased 1.4 °C between 1951 and 2000 (Hartmann and Wendler, 2005). Along with this so-called "polar amplification" of warming (Glossary), rapid melting of sea ice, stress on wildlife populations, increased fire frequency and severity, and other iconic changes have taken place in the Arctic in recent decades (see references above).

Boreal forests are undergoing substantial biome shifts as a result of these rapid changes (Barber et al., 2009; Andreu-Hayles et al., 2011; Beck et al., 2011; Beck et al., 2013; Juday, 2011; Xu et al., 2013). This is particularly noteworthy because boreal forests, estimated to account for approximately 22% of the carbon stored in forests worldwide, play an essential role in the carbon balance of the globe (Pan et al., 2011; Milakovsky et al., 2012). They also, importantly, represent some of the last remaining undisturbed wilderness areas on the earth today (ACIA, 2005). There is ample evidence that such shifts in growth patterns,

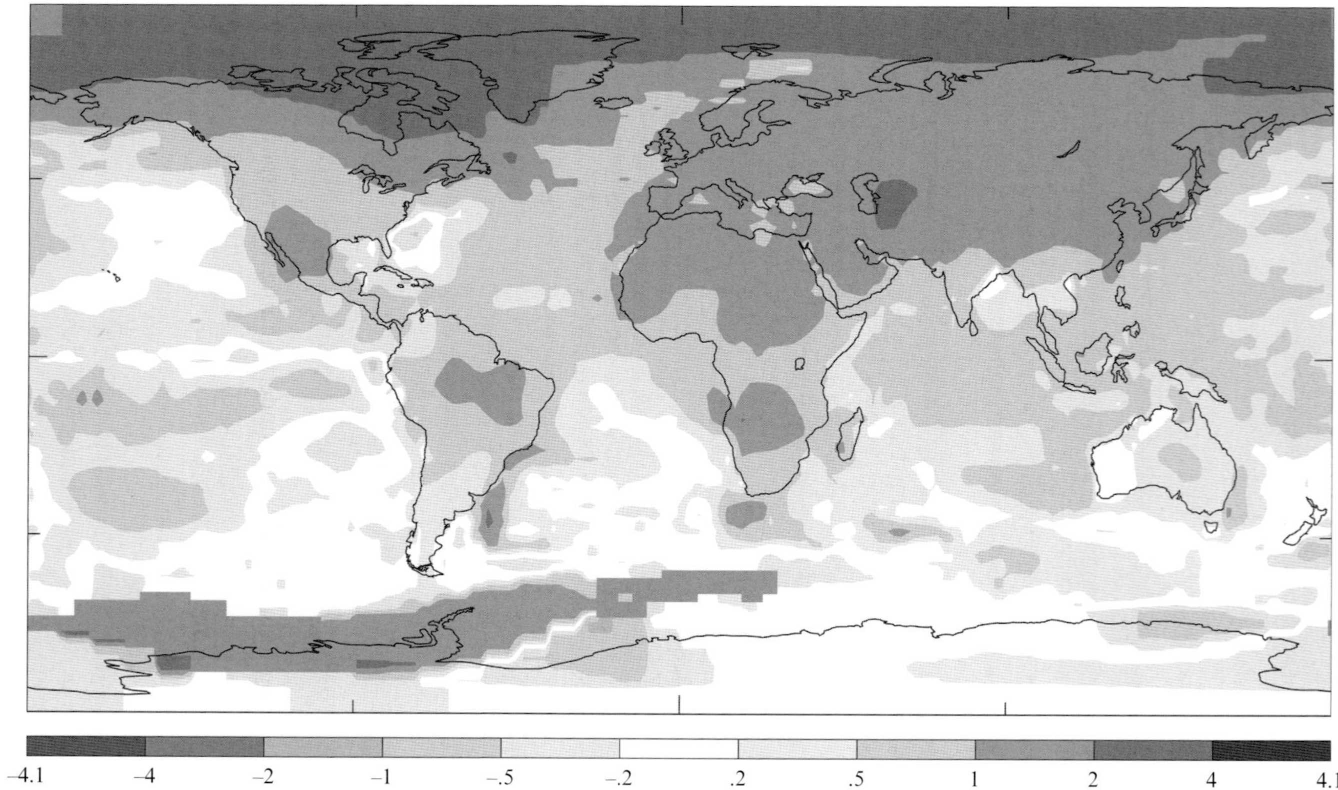

Figure 1.2 Map showing evidence from annual (Jan–Dec) temperature data of large-scale warming and polar amplification in recent decades (temperatures averaged over 1975–2010 in degrees Celsius). Adapted from Goddard Institute for Space Studies (GISS) (http://data.giss.nasa.gov/gistemp/maps/; Hansen et al., 2010). For color detail, please see color plate section.

such as migration of treeline position (Esper and Schweingruber, 2004), earlier growing season onset, and declining trends in forest productivity, are well under way, based on both satellite vegetation monitoring and ground-based observations (Bunn and Goetz, 2006; Beck et al., 2011; Beck et al., 2013; Juday, 2011).

The response of northern forest ecosystems to change appears to be complex and possibly nonlinear. This response can, in turn, cause significant feedbacks into the climate system, including changes to the planet's albedo and carbon budget (Chapin et al., 2004; Jansen et al., 2007 [IPCC]; Lloyd and Bunn, 2007; Bonan, 2008). Widespread tree mortality in interior Alaska due to drought stress, fire, and insect outbreaks has been attributed to the negative impacts of global warming on forest growth (Juday, 2011). A related possible response is the observed decoupling between tree growth and temperature data, commonly now referred to as "divergence," observed mainly in northern latitude treeline settings (e.g., D'Arrigo et al., 2008; Glossary; see later chapter for a discussion). It is important to note, however, that many northern forests are still responding positively to warmer temperatures, recording this signal in the recent period as well as back in time (e.g., D'Arrigo et al., 2001; Wilson et al., 2007b; Grudd, 2008; Esper et al., 2010; Beck et al., 2011; Büntgen et al., 2011; Melvin et al., 2012).

The tree-ring data and chronologies described herein have been transformed into climatic (temperature) reconstructions (Glossary), which are defined as records of local, regional to larger-scale thermal histories that extend back prior to the available instrumental record. These reconstructions are primarily related to temperature but also reflect, and integrate, to varying degrees, past fluctuations in hydroclimate, sea-level pressure, and synoptic and broad-scale atmosphere-ocean circulation dynamics and teleconnections. The tree-ring data we describe also yield valuable information and observations on forest growth productivity and how it has varied over annual to centennial time scales.

The generation of large-scale reconstructions of temperature from tree rings and other proxy archives is an ongoing, evolving process. These records will continue to improve as additional data coverage and new methodologies become available. Such data archives are essential for (1) constraining the sensitivity of the Earth's climate system, (2) placing recent anthropogenically-forced variability in a long-term context, (3) understanding the response to both natural external radiative and other forcings (solar, volcanic, greenhouse gases, aerosols) and internal variability (e.g., the El Niño-Southern Oscillation or ENSO, the North Atlantic Oscillation or NAO, Arctic Oscillation or AO; all in Glossary), and (4) facilitating and providing input for constraining future modeled scenarios derived from climate and vegetation models via proxy/model comparisons. At the same time, this information must take into consideration the rapid, and at times unprecedented changes in forest growth and temperature now taking place.

1.4. "Northern Archive" Synthesis

We synthesize scientific results outlined in more than 50 peer-reviewed papers published over the past four decades by TRL-LDEO investigators on northern forests, in addition to various proceedings volumes and reports (e.g., National Research Council, National Academy of Sciences, Intergovernmental Panel on Climatic Change or IPCC). A listing of these core papers on northern forests over the past few decades is provided near the end of this monograph. Also included is an annotated compilation of long-term tree-ring raw data measurements, metadata, chronologies, and reconstructions, which is being made available under the heading "Northern Archive" on the NOAA Paleoclimatology (www.ncdc.noaa.gov) website. This data is being made available via the Internet for scientists, teachers, students and the public to allow them to pose their own questions about the research and to use the data for their own efforts.

2

2. TREE-RING INVESTIGATIONS AT NORTHERN LATITUDES

2.1. Initial Studies

During the development of the science of dendrochronology in the early twentieth century (Fritts, 1976), tree-ring chronologies from the southwestern United States were used to reconstruct indices of precipitation, temperature, and other variables of local (individual sites) to regional (small networks of sites, covering~synoptic-scale or larger) extent, based on wood sampled from old growth forests in relatively wild yet readily accessible regions. More recent studies have used even larger-scale tree-ring record networks across North America to reconstruct past drought variability for the past millennium (e.g., Cook et al., 2004; Cook et al., 2010).

Up until the middle of the twentieth century, however, old growth forests at the northern North American treeline had scarcely been investigated from a tree-ring perspective due to the remote nature of these wilderness locations. This is in contrast to the northern treeline of Eurasia, where such early studies were more common due to easier access to many forest sites (e.g., Schulman, 1944).

One of the pioneers of northern North American tree-ring research was James Louis Giddings (1909–1964), a largely self-taught dendrochronologist and archaeologist who demonstrated the potential for Arctic dendrochronology (Nash, 2000). Giddings published a number of Arctic studies describing tree-ring dating of archaeological dwellings, structural timbers, and other wood material, mainly of white spruce (*Picea glauca* [Moench] Voss), the dominant western North American treeline species. His seminal work was the generation of a tree-ring width chronology for the Kobuk-Noatak-Selawik River drainages of northwestern Alaska, derived from wood collected over a 24,000-km^2 region of the Noatak and Kobuk river basins that dated back to 978 AD (e.g., Giddings, 1941). Other Arctic dendrochronologists, including Wendell Oswalt and James Van Stone (see references in Nash, 2000), also conducted early investigations. Efforts to collect archaeological and other ancient wood material are still ongoing in areas of the far north, in locations

---

Dendroclimatic Studies: Tree Growth and Climate Change in Northern Forests,
Monograph Number 67. Rosanne D'Arrigo, Nicole Davi, Gordon Jacoby,
Rob Wilson, and Greg Wiles.

where subfossil wood (i.e., wood that has been dead for several hundreds to thousands of years, Glossary) have been well preserved.

The TRL-LDEO was first established in the 1970s (see chapters by Jacoby and D'Arrigo, in Lippsett, 1999), founded by Jacoby and Cook. D'Arrigo came to the TRL in the early 1980s as a graduate student and then as a research scientist. Building on the original work of Giddings (1941), Jacoby, with Cook and later with D'Arrigo and other investigators, including Davi and Wiles, helped expand upon these early tree-ring studies of the 1970s, along the latitudinal treeline of North America and other locations (see Figure 1.1). More recently, Wilson of the University of St. Andrews, UK (and Adjunct Research Scientist, LDEO), has contributed extensive data processing, analysis, interpretation, and publications to this ongoing research (see Figure 1.1).

Initially, TRL-LDEO-related efforts targeted remote treeline areas of Alaska and Canada's Yukon Territory, eventually expanding southward to the coastal forests of the Gulf of Alaska region, and eastward to other Canadian provinces, including the Northwest Territories, Nunavut, Quebec, Labrador-Newfoundland, and the Canadian Maritimes (see acknowledgments for listing of other significant contributors to and collaborators of this work). Across Alaska and northern Canada, the tree-ring collections have yielded annually resolved paleoclimatic time series that integrate temperature-related conditions, as well as related aspects of climate and atmospheric circulation and teleconnections for the region.

This began a series of research projects on tree-ring studies of northern forests, largely funded by the National Science Foundation, the National Oceanic and Atmospheric Association (NOAA), and other agencies, which have now spanned nearly four decades. As noted, these archives have been utilized primarily to extend the relatively short instrumental record of climate for these northern latitude regions, but also to derive information on the response of this boreal forest component of the terrestrial biosphere to climatic and environmental change in the anthropogenic era and in the past.

2.2. Site Selection

In selecting northern sites for sampling, we use observational criteria that aid identification of trees with sensitivity to past temperature variability, that minimize complications from non-climatic factors, and that are indicative of advanced age. Such criteria include the absence of any obvious disturbance (e.g., due to fire, insect infestation, storm damage); selection of trees that are widely spaced rather than closed canopy to minimize any complicating effects of stand dynamics; presence of mesic ground cover, which indicates that drought stress is likely to be minimal; and presence of heavy lower branches, dead crowns, and stunted tree morphology, which are indicators of age and sensitivity to environmental stress (Jacoby and D'Arrigo, 1989). Another important strategy or philosophy that we advocate is to select tree sites and chronologies that have strong common variability at a range of frequencies (interannual up to centennial). Site location is critical in this regard, because it is mainly the most undisturbed, open canopy treeline locations that yield such coherent low-frequency climatic information. However, it is important to note that not all tree-ring sites from which samples are collected will necessarily be found to retain such low-frequency information, and thus be deemed appropriate for such dendroclimatological studies.

2.3. Tree-Ring Parameters and Processing: Ring Width and Maximum Latewood Density

Annual tree-ring data are transformed into final time series or chronologies after being rigorously cross-dated, measured, and processed to retain as much potential low-frequency climate information as possible (Stokes and Smiley, 1968; Fritts, 1976; Cook, 1985; Cook and Kairiukstis, 1990; Briffa et al., 1992; Cook et al., 1995). The resulting records of radial growth, or dimensionless tree-ring chronologies, have been used to generate dendroclimatic reconstructions of temperature (e.g., Esper et al., 2002; D'Arrigo et al., 2006; Wilson et al., 2007b) and related variables (e.g., freeze dates for Hudson's Bay, Jacoby and Ulan, 1982) for individual sites and regional locations in northern forests.

Both more traditional (e.g., negative exponential, straight line curve fitting) and more recently applied, innovative methods such as Regional Curve Standardization (RCS, Glossary) have been employed for detrending tree-ring time series (Melvin and Briffa, 2010). Such detrending methods of standardization (Glossary) are used to estimate and remove age-related (biological) growth trends and end effects (Glossary) prior to generation of tree-ring chronologies (Cook and Kairiukstis, 1990). RCS allows the capture of low-frequency, even centennial trends that can exceed the length of individual tree samples used in chronology development, but this method may not be applicable to all sites (see Cook et al., 1995 and the 'segment length curse,' Esper et al., 2003). Alternative techniques, such as age-band decomposition (ABD; Glossary) (Briffa et al., 2001), and signal-free standardization methods (Melvin and Briffa, 2010) can also be employed to minimize potential biases in RCS-related methodologies. Such methods are under continual refinement, and such research is critically important if we are to resolve issues in our understanding of low-frequency variability of climate (Fritts, 1976; Cook and Kairiukstis, 1990; Esper et al., 2003; Esper et al., 2004; Melvin, 2004; Melvin and Briffa, 2008; Briffa and Melvin, 2010).

Most dendroclimatological studies utilize two tree-ring parameters: ring-width and maximum latewood density (Glossary). Ring widths are by far the most commonly used parameter for tree-ring analyses. Ring-width chronologies have been shown to be sensitive to summer temperature conditions at northern treeline sites, but they can also integrate temperatures throughout the year (see discussion below on Northern Hemisphere [NH] reconstructions) (Jacoby and D'Arrigo, 1989; Fritts, 1976; Schweingruber, 1988; Cook and Kairiukstis, 1990; Osborn and Briffa, 2006). Ring widths from temperature-sensitive coniferous treeline sites (e.g., white spruce) typically feature strong autocorrelation (Glossary) effects that tend to also smooth out response to some high-frequency climatic or environmental events (Anchukaitis et al., 2012a; D'Arrigo et al., in press).

In addition to ring-width tree-ring series, the TRL-LDEO was one of the first laboratories to develop the capability to generate records of maximum latewood density (MXD), a parameter that is often an excellent indicator of extended warm season annual to multidecadal or longer temperature variability. MXD is formed in the latter part of the growing season and is defined as the peak density value of the last formed latewood cells measured in an annual growth ring (Schweingruber, 1988; D'Arrigo et al., 1992). Physiological fluctuations in wood density over time are related to the activity of the vascular cambium (Glossary) and vary with the season, age, climate, and environmental conditions (e.g., Schweingruber, 1988). Visually, latewood consists of thicker-walled cells with smaller lumen, which creates a visual contrast with the earlywood cells of the same ring (gradual transition) and those formed in the subsequent spring season (abrupt transition). We (see also Thetford et al., 1991) have employed the technique of X-ray densitometry (Glossary), with image analysis software using the Dendro-2003 (Walesch) system. This method identifies ring and intra-ring boundaries in the wood to generate density time series, including maximum latewood density, minimum earlywood density, and other intra-ring parameters.

The climate signal derived from MXD (and related indicators, such as earlywood and latewood width and earlywood density) complements the summer, and annual, multidecadal-to-centennial scale temperature information that can often be derived from ring-width data at climatically sensitive northern sites (Schweingruber, 1988; Jacoby and D'Arrigo, 1989; Davi et al., 2003; Wilson and Luckman, 2003; D'Arrigo et al., 2006). Typically, MXD series from boreal conifers correlate positively with monthly temperatures over an extended summer season (e.g., from May through August or September), making it an excellent proxy for warm-season temperature and extreme high-frequency events (Fritts, 1976; Schweingruber, 1988; Briffa et al., 1992; D'Arrigo et al., 1992). In some cases, centennial or longer time-scale information can be gleaned from density archives (e.g., Luckman and Wilson, 2005; Büntgen et al., 2006; Grudd, 2008; Esper et al., 2012).

Although we have used both ring-width and density series in our analyses, our large-scale temperature reconstruction efforts to date (Jacoby and D'Arrigo, 1989; D'Arrigo et al., 2006), have mainly been based on ring widths. This is because density processing is very labor intensive, with greater attrition of wood samples. Thus, fewer of these records have been generated overall, although there is much potential to develop density series for already-sampled sites processed for ring width, given adequate resources.

3

3. SELECTED LOCAL TO REGIONAL TRL-LDEO NORTHERN TREE-RING STUDIES

One of the first northern treeline studies by TRL-LDEO investigators published for the far north described the tree-ring width chronology named Twisted Tree-Heartrot Hill (TTHH), from a white spruce site in the Yukon Territory (YT), Canada (Jacoby and Cook, 1981) (Figure 3.1).

The trees at the TTHH site exhibit morphological features and microsite conditions that were deemed consistent with a tendency for a strong temperature response. This site is the last stand of trees, found at elevational treeline, along a mountain pass of the Dempster Highway, YT. The ring-width chronology is significantly correlated with summer temperatures at the nearby Dawson, YT meteorological station, and this relationship was used to infer information on summer temperatures over the past 400 years (Jacoby and Cook, 1981). Notably, in addition to temperature sensitivity, there was some evidence, even at the time of this study several decades ago, of late summer drought stress (see below).

Growth trends at this treeline site demonstrated cooling (as inferred by narrow ring widths) during the very late 1600s, followed by warming in the 1700s. There was pronounced cold inferred during the early 1800s, one of the peak cold periods of the so-called Little Ice Age (LIA, Glossary) interval, followed by a recovery and inferred warming in the later 1800s. Even greater warmth is inferred for the subsequent anthropogenic era of increasing greenhouse gases (but with declining trends in recent decades, see below). These trends are broadly evident at other circumpolar northern treeline sites, as well as some elevational treeline sites at lower northern latitudes (e.g., D'Arrigo et al., 2001, Mongolia; Büntgen et al., 2011, Europe), and generally reflect large-scale, low-frequency temperature fluctuations over the past millennium. Generation of such site chronologies as TTHH, and development of local to regional temperature reconstructions, helped set the stage for the larger-scale reconstruction efforts outlined below.

Dendroclimatic Studies: Tree Growth and Climate Change in Northern Forests,
Monograph Number 67. Rosanne D'Arrigo, Nicole Davi, Gordon Jacoby,
Rob Wilson, and Greg Wiles.

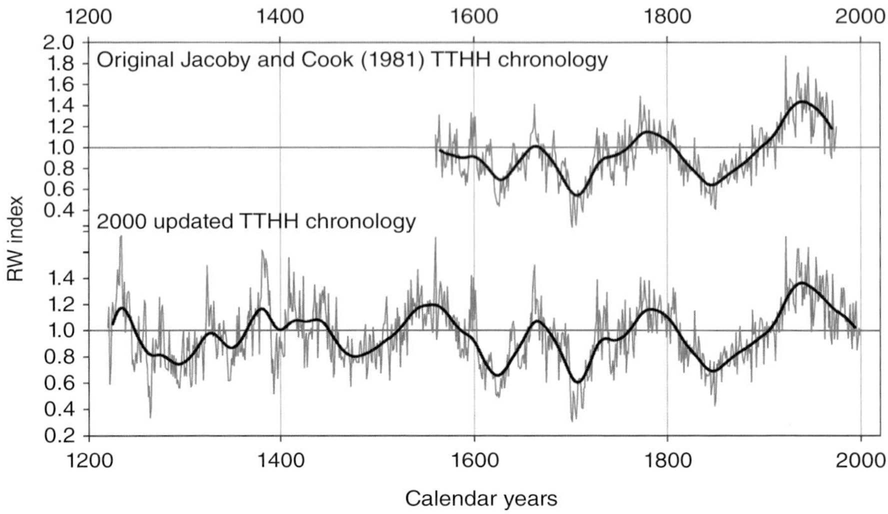

Figure 3.1 Original (*top*) and updated Twisted Tree-Heartrot Hill (TTHH) tree-ring width chronologies from elevational treeline, Yukon Territory, Canada. Units are in dimensionless indices. Adapted from Jacoby and Cook (1981). For color detail, please see color plate section.

Several millennial to submillennial-length tree-ring chronologies have recently been generated (2010–2011) for northern Alaska, representing some of the few such records of this length for northwestern North America. These include latitudinal treeline ring-width and MXD records for the Firth River area, in the Arctic National Wildlife Refuge (ANWR) of northeastern Alaska (Anchukaitis et al., 2012b). Anchukaitis and colleagues (2012b) developed a reconstruction based on a stable relationship between Firth River MXD and regional summer temperatures. The warmest epoch during the past 900 years occurred during the twentieth century, with multidecadal trends prior to that time co-varying with changes in radiative forcing due to solar and volcanic activity (Anchukaitis et al., 2012b). Another millennial-length chronology at Mount Sukakpak, near the Dalton Highway, is based on living and subfossil wood, which was first sampled in the 1970s. The response of the trees to climate at this location is complex. In addition to temperature, they are also sensitive to drought, due to relatively dry microsite conditions at this site.

In northwestern Alaska, the archaeological wood samples collected by Giddings (1941) from old dwellings and other sources (see also Graumlich and King, 1997) were obtained by TRL-LDEO scientists and combined with additional samples from the Seward Peninsula, Alaska, to develop long composite ring-width and density chronologies, which were linked to features of large-scale North Pacific climate variability (e.g., coastal land and ocean temperatures, and circulation indices such as the Pacific Decadal Oscillation or PDO) (Mantua et al., 1997; D'Arrigo et al., 2005a; D'Arrigo et al., 2005b).

Signatures of North Pacific climate variability, such as the PDO, and the notable warming associated with the Pacific regime shift of 1976–1977 (Mantua et al., 1997) are most strongly reflected in millennial and submillennial records that we have developed from the Gulf of Alaska region (Barclay et al., 1999; Wiles et al., 2004; Wilson et al., 2007a), based on coastal species of spruce, hemlock, and other long-lived conifers. These records, dating back to the first millennium AD, have also yielded important information about past glacial advance and retreat (Figure 3.2) (Wiles et al., 2004; Wiles et al., 2008; Wiles et al., 2011).

In another regional-scale Alaskan study, Jacoby and colleagues (1999) compiled a small network of maximum latewood density chronologies to reconstruct summer temperatures for interior Alaska. The lowest reconstructed temperature value over this 400-year record was found to occur in the summer of 1783, following the eruption of Laki, Iceland that began in June of that year (Figure 3.3).

Inuit legends describe widespread winter-like conditions and loss of life due to famine during that fateful summer (Oquilluk, 1973). Maximum latewood density, due to its strong response to summer temperatures, low autocorrelation relative to ring width, and apparent sensitivity to fluctuations in solar radiation (e.g., Esper et al., 2012), is a particularly powerful parameter for studying extreme cold years that coincide with volcanic events (Briffa et al., 1998a; Anchukaitis et al., 2012a; D'Arrigo et al., in press).

In central Canada, we have described additional tree-ring records for wilderness areas at latitudinal treeline near the Coppermine (now called Kugluktuk) River, Nunavut and Northwest Territories, and the Thelon River Sanctuary, Nunavut (see core TRL-LDEO papers, e.g., D'Arrigo et al., 2009; Andreu-Hayles et al., 2011). The Coppermine ring-width chronology dates back to 1046 and is one of the very few such long series for Canada. For both the Coppermine and Thelon sites, the ring-width data show a greater decoupling from recent temperature trends than does the corresponding latewood density data. The latter also shows an increased sensitivity to drought over recent decades, supporting the observation that temperature-induced drought stress may be a contributing factor to divergence at these particular locations.

Tree-ring records generated by the TRL-LDEO have also been developed for a number of sites along the Eurasian boreal forest treeline and adjacent temperature-sensitive locations. One such location is the Taymir Peninsula, Siberia, where several chronologies were generated from the northernmost forests on earth (Jacoby et al., 2000). These records were used to reconstruct May to September temperatures for the region, but were shown to lose their thermal response with local station temperature records after around 1970.

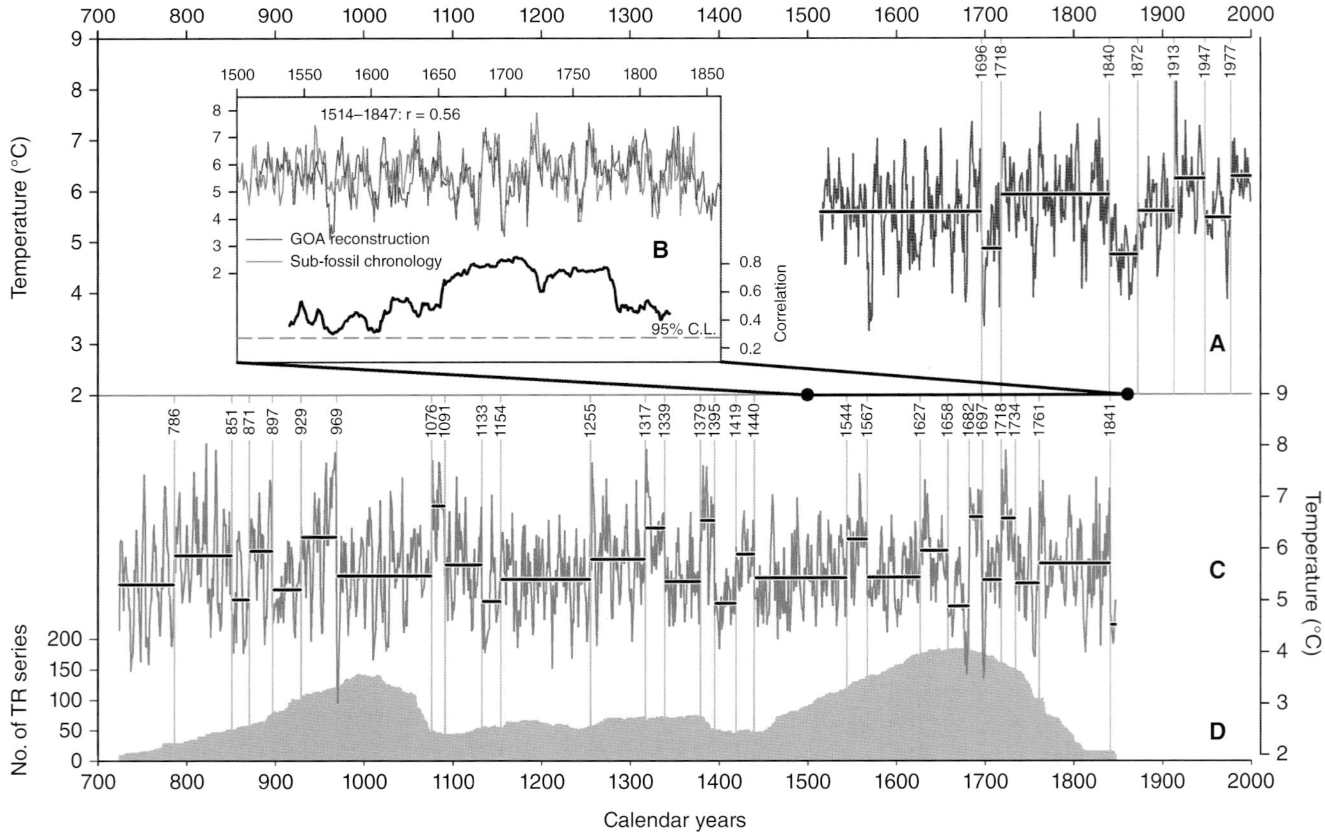

Figure 3.2 Comparison of Gulf of Alaska (GOA) living and subfossil tree-ring width chronologies. The subfossil data have been scaled to the reconstruction over the period of overlap (1514–1847). **A**. The GOA reconstruction time series. **B**. Period of overlap with sliding 31-year correlations. **C**. The scaled subfossil chronology. The vertical lines (with dates) in **A** and **C** show significant (95% confidence level) regime shifts in the time series interpreted as representing decadal variability in North Pacific climate. **D**. Series replication in the subfossil chronology. Adapted from Wilson et al. (2007a). For color detail, please see color plate section.

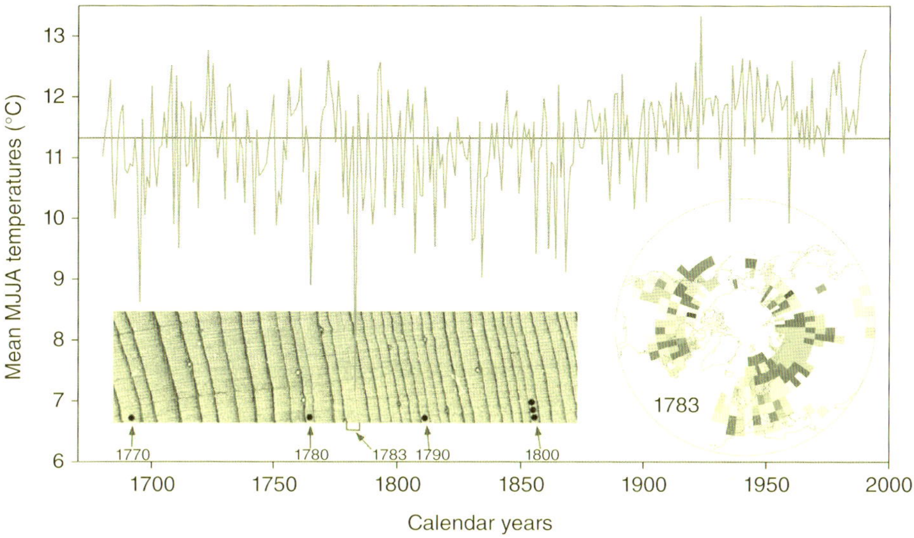

Figure 3.3 Tree-ring reconstruction of May–August temperatures for Alaska based on maximum latewood density data. Note anomalously low reconstructed temperatures during 1783, the year of the major Laki volcanic eruption in Iceland. Adapted from Jacoby et al. (1999). For color detail, please see color plate section.

Other intensive TRL-LDEO efforts in Eurasia include a long-term project in Mongolia (termed "MATRIP"), which has been ongoing since 1995 (Jacoby et al., 1996; D'Arrigo et al., 2001; Pederson et al., 2001; Davi et al., 2010). There now exists fair coverage of centennial to millennial length tree-ring records for Mongolia, which have been used to reconstruct indices of temperature (Jacoby et al., 1996; D'Arrigo et al., 2000; D'Arrigo et al., 2001a) as well as precipitation and related variables (Pederson et al., 2001; Davi et al., 2006; Davi et al., 2010).

For example, a temperature-sensitive elevational treeline record from Sol Dav, Mongolia dates back two millennia and revealed a positive response to warming observed over central Asia in recent decades (Jacoby et al., 1996; D'Arrigo et al., 2001). Notably, a narrow growth ring with frost damage (Glossary) in 536 AD was attributed to a major, unknown volcanic event at that time. This volcanic event has also been described in several other long tree-ring chronologies from sites around the globe (Baillie, 1999; D'Arrigo et al., 2001b and references cited therein; Larsen et al., 2008).

A long-term TRL-LDEO research project in the western North Pacific has yielded tree-ring records for the Kurile Islands, Kamchatka, and the Russian Far East, and Japan (D'Arrigo et al., 1997; Davi et al., 2002; Jacoby et al., 2004; Solomina et al., 2006). Jacoby and colleagues (2004) describe the development of a 400-year reconstruction of temperature based on oak (*Quercus crispula*) trees that has a strong relationship with the PDO index, a large-scale indicator of temperature and atmosphere-ocean climate variability for the North Pacific sector. This finding indicates the potential for tree-ring records from this region to be useful in improving reconstructions of the PDO and related indices, to expand upon prior versions based solely on western North American tree-ring data (e.g., Biondi et al., 2001; D'Arrigo et al., 2001; Villalba et al., 2011). There is, in fact, increasing evidence that the climate of northeastern Asia plays a critical role in the northern Pacific climate and its teleconnections to North America and other locations, although the typical response to the PDO is stronger over North America (D'Arrigo and Wilson, 2006; Shen et al., 2006).

4

4. THE BROADER CONTEXT OF NORTHERN DENDROCLIMATIC STUDIES

In this section, we briefly outline the broader scope of dendroclimatic research that has been published to date for far northern areas. This overview is not meant to be at all comprehensive, and only aims to provide a glimpse into the broader context for the TRL-LDEO research described herein. A true synthesis that will bring these various datasets generated by various laboratories and researchers together has not yet been done and could be a useful focus for future research on the boreal regions of North America and Eurasia.

4.1. North America

Tree-ring research in Alaska and adjacent northwestern Canada by Juday (2011) and colleagues (e.g., Barber et al., 2000; Wilmking et al., 2005) using ring-width, MXD, and isotopic records has focused on the relatively closed canopy interior boreal forests, including those in the Bonanza Creek watershed near Fairbanks. Their analyses have clearly demonstrated that the unprecedented, rapid warming in Alaska over recent decades (ACIA, 2005), along with coincident temperature-induced drought stress and intensified insect activity, have resulted in large-scale tree mortality and growth declines across the region. In broader transects across much of Alaska, Wilmking and colleagues (2005) demonstrated that drier sites show a higher percentage of trees that are responding negatively to temperature in recent decades, due to greater drought stress and increased evapotranspiration. This work is consistent with evidence of divergence effects elsewhere in Alaska and the far north (e.g., Jacoby and D'Arrigo, 1995; see below).

Tree-ring records for the Seward Peninsula in Alaska, and elsewhere in the circumpolar north, were generated by Lloyd and Fastie (2002 and 2003), and

Dendroclimatic Studies: Tree Growth and Climate Change in Northern Forests,
Monograph Number 67. Rosanne D'Arrigo, Nicole Davi, Gordon Jacoby,
Rob Wilson, and Greg Wiles.

Lloyd and Bunn (2007), which complement the TRL-LDEO research in the Seward Peninsula and northwestern Alaska, as noted above. These authors investigated the role of shifting seasonality of warm-season precipitation as a key driver of forest growth patterns and fire regimes in the modern anthropogenic era relative to the past.

In the Canadian Rockies and other locations, Brian Luckman (University of Western Ontario) and colleagues have generated long tree-ring chronologies for use in climatic reconstructions (e.g., Youngblut and Luckman, 2008; Luckman 2010). Luckman, along with R. Wilson, produced millennial-length ring-width and density series for the Icefields region of Alberta (extending back to AD 950), which were combined to reconstruct summer (May to August) maximum temperatures (Luckman and Wilson, 2005). Linkages were identified between this reconstruction and periods of glacial advance as well as impacts of solar and volcanic activity (Luckman and Wilson, 2005). This Icefields record was included in the D'Arrigo and colleagues (2006) NH temperature reconstruction, and shows coherent trends with other series used in this compilation for the past millennium.

Also in western Canada, Michael Pisaric and Trevor Porter of Carleton University, Ottawa, have developed isotopic and other tree-ring parameter records for the Mackenzie River Delta, Northwest Territories (NWT) in order to study long-term changes in precipitation patterns, and have reconstructed early-season precipitation and landscape conditions for the Yellowknife, NWT area using jack pine and other species (Pisaric et al., 2007; Porter et al., 2009; Porter and Pisaric, 2011; Porter et al., 2013). They also employed a 23-site network of white spruce ring-width chronologies from near treeline in the Yukon Territory that shows evidence of warming-induced drought stress, as well as growth declines that depend upon site-level factors (see divergence section below).

S. Payette and L. Filion of the University of Laval, Quebec, have long studied the ecology (fire, wind, and snow activity) and past climate of old growth forests in eastern Canada (Payette and Filion, 2011). Their research included the study of so-called "light density rings" (Glossary) that were associated with past volcanic events such as that of Tambora's eruption in 1815, and the subsequent 1816 "year without a summer". Their research group has developed millennial ring-width and density records based on living and subfossil black spruce (*Picea mariana* [Mill]), a common tree species in eastern Canada (Wang et al., 2001).

The Labrador Highlands Research Group of Memorial University, Newfoundland, Canada (http://www.mun.ca/geog/lhrg/) has used tree rings to reconstruct regional patterns of climate and forest treeline ecology and migration for northeastern Canada, including Labrador and Newfoundland, and relationships with local populations and culture (e.g., Banfield and Jacobs, 1998; Labrador Highlands Research Group, 2004).

4.2. Eurasia

Across Eurasia, dendroclimatic research in temperature-sensitive locations has included numerous studies, for example, in Scandinavia (Briffa et al., 1992; Timonen, 2007; Melvin et al., 2012; Esper et al., 2012), the Alps (Frank and Esper, 2005; Büntgen et al., 2006), the United Kingdom (Wilson et al., 2011), and Russia, including summer temperature-sensitive records in the Polar Urals (Esper et al., 2002), Yamal (Hantemirov and Shiyatov, 2002), Taymir (Naurzbaev and Vaganov, 1999), the Altai Mountains (Panyushkina et al., 2005), and Yakutia (Hughes et al., 1999). See also Solomina and Alverson's (2004) review and references therein.

4.3. Tree-Ring Chronology Networks

In the 1980s, several major field expeditions were conducted along large-scale transects across the northern forests to collect samples of wood for dendroclimatic analyses. The Schweingruber transect (http://lwf.ncdc.noaa.gov/paleo/treering-wsl.html with G. Jacoby) was one of these, a joint project between WSL, Switzerland, and the TRL-LDEO. Samples were collected in the summer of 1982 from boreal forest sites at or near elevational or latitudinal treeline, and processed for both ring-width and density chronologies. Another latitudinal tree-ring transect covered western North America and also yielded a multiple ring-width and density network (Briffa et al., 2001). These networks, along with similar networks and datasets from across the Eurasian Boreal region, have been used in a number of studies to investigate large-scale patterns and trends of forest growth and climatic change (e.g., Briffa et al., 1998a; Briffa et al., 1998b; Briffa et al., 2001).

The chronologies developed by TRL-LDEO scientists and colleagues for northern locations have also been compiled into a circumpolar network, based on data from latitudinal treeline sites that have been used to generate larger-scale temperature reconstructions.

5

5. TEMPERATURE RECONSTRUCTIONS FOR THE NORTHERN HEMISPHERE

5.1. Initial Attempts

The science of generating large-scale dendroclimatic reconstructions has continued to evolve as spatial coverage of sites and availability of reconstruction methodologies have improved over the past several decades. High-resolution, near-hemispheric-scale temperature reconstructions only became possible once proxy coverage had expanded across the northern treeline, allowing the development of reconstructions beyond local to regional scales that could be used to infer large-scale climatic information with an emphasis on lower frequencies.

Groveman and Landsberg (1979; and see review in Frank et al., 2010) were the first to publish a quantitative, multiproxy reconstruction of annual Northern Hemisphere (NH) temperatures. Their record was based just partly on tree rings—only 2 series out of 20—the rest being long instrumental records or other proxy time series. Their reconstruction was developed using simple nested multiple regression models (specifying time-varying estimates of uncertainty) dating back to 1579. Their study was the first to demonstrate that a few carefully selected series could provide a reasonable estimation of temperatures for the NH.

Another pioneering study in this regard was the publication of a reconstruction of NH annual temperatures for 1671–1973, by Jacoby and D'Arrigo of the TRL-LDEO (1989, in the journal *Climatic Change*). It was based on a network of just 11 climatically sensitive tree-ring-width chronologies from northern North America, and was the first to reconstruct temperature variability for the NH solely from tree rings. Although the number of sites was not large, and the sites were biased to North America, their coverage represented a substantial expanse of the northern forest landscape, from Alaska to easternmost Canada. Jacoby and D'Arrigo (1989) argued that this spatial representation was adequate for this initial study, particularly given the representativeness of higher latitude North American temperature patterns to the hemisphere as a whole.

Dendroclimatic Studies: Tree Growth and Climate Change in Northern Forests,
Monograph Number 67. Rosanne D'Arrigo, Nicole Davi, Gordon Jacoby,
Rob Wilson, and Greg Wiles.

This record was used to assess past temperature variability and the degree to which recent anthropogenic warming had been unusual relative to past centuries (Jacoby and D'Arrigo, 1989). The initial dataset was substantially expanded and updated to encompass sites in Eurasia as well (D'Arrigo et al., 2006), and to derive temperature estimates for the full millennium.

The Jacoby and D'Arrigo (1989) reconstruction used principal components analysis and linear regression techniques, and provided statistical calibration and verification analyses of regression modeling results. One key finding of this study was that the warmth of the twentieth century was unusual relative to the previous several hundred years. This early work on reconstructing NH temperatures, and our observations of treeline forest growth in response to warming, were presented at one of the first-ever hearings of the U.S. Senate on global warming and climatic change in 1992, chaired by then Senator Albert Gore.

From a historical standpoint, it is worth noting that there was considerable resistance at the time of the 1989 study from the dendrochronological and climate communities to the idea of reconstructing hemispheric-scale temperatures from a select small number of northern treeline sites, and disagreement as to what level of site coverage could be considered adequate for reflecting such large-scale variability. Another topic, still controversial, has been whether or not ring-width data, composited at large hemispheric scales, are accurate indicators solely of *summer* temperatures, or whether ring widths might also, under certain conditions, integrate *annual*, large-scale, low-frequency climate variability. All of these issues are still topics of discussion and debate today.

We have long interpreted the response of northern treeline forests to the thermal regime to occur integratively, with response to air temperature, solar radiation, and soil temperatures all playing roles (Jacoby and D'Arrigo, 1989). Although cambial cell division in the far north can be limited to only about six weeks, other growth aspects (e.g., photosynthesis, root extension) can occur over a much longer period. For example, northern trees can respond positively to prior fall temperatures, when production and storage of carbohydrates take place for the following year's growth (e.g., Tranquillini, 1979; Havranek and Tranquillini, 1995). Also, milder winters can reduce the amount of heat needed to thaw root zones and active layers in areas of continuous and discontinuous permafrost (Jacoby and D'Arrigo, 1989, and references cited therein, e.g., Kozlowski, 1971; Tranquillini, 1979; Goldstein, 1981). These northern trees have the potential to store nutrients in the fall for the next year's growing season, and can benefit from warmer early spring conditions provided that adequate moisture is available. There are thus a number of reasons to support the argument that trees can integrate conditions beyond the actual summer growing season, although summer is still considered the optimal target for many paleotemperature studies (e.g., Wilson et al., 2007b).

5.2. Evolution of NH Temperature Reconstructions

Jacoby and D'Arrigo (1989 and subsequent updates: D'Arrigo et al., 1993; D'Arrigo et al., 1999; D'Arrigo et al., 2006) described their study as a first iteration in an ongoing process by which such large-scale climate paleoestimates would continue to improve over time. This has since proven to be the case, with a considerable number of additional reconstructions published in the past two decades, spurred on by the emergence of climate change science, and improved proxy and instrumental data coverage and techniques. The TRL-LDEO's northern tree-ring data archive has served as a baseline for more detailed and complex reconstructions of NH temperature series, mainly since the late 1990s, some extending back to the past millennium (e.g., Mann et al., 1998; Mann et al., 1999 [the so-called "hockey stick" reconstructions], Jones et al., 1998; Esper et al., 2002; Moberg et al., 2005; Hegerl et al., 2006; Osborn and Briffa, 2006; Hegerl et al., 2007; Mann et al., 2008; Christiansen et al., 2009). The last few decades have seen a gradual transition in the field of dendroclimatology from single and regional site studies to multi-network analyses that are actively being used to improve estimates of large-scale climatic processes of the Earth's climate system for the past millennium. There has thus been considerable evolution and progress since the "Eurocentric" cartoon depicted in the 1990 IPCC report, believed to represent temperature changes over central England, and on a larger scale over the past 1,000 years (Frank et al., 2010).

The reports of the IPCC (e.g., Jansen et al., 2007) outline how the science of large-scale temperature reconstructions has evolved from the simpler early models (e.g., Groveman and Landsberg, 1979; Jacoby and D'Arrigo, 1989) to the relatively more complex reconstructions of the past decade or so (e.g., Mann et al., 1998; Mann et al., 1999; Mann et al., 2008; Mann et al., 2009; Rutherford et al., 2005; Frank et al., 2010; Christiansen and Ljungqvist, 2012). The most recent IPCC report (2007) illustrates how the increase in uncertainties back in time among the various reconstructions precludes identification and consensus of the precise amplitude of past temperature changes (Esper et al., 2005; Frank et al., 2010).

An update on our original 1989 NH temperature reconstruction, published in 2006 (Tables 5.1 and 5.2, Figures 5.1–5.4) used original and recently sampled tree-ring records from North America and Eurasia to reconstruct annual extratropical NH temperatures for the past millennium (D'Arrigo et al., 2006). Tree-ring records were only included if they passed an initial criterion of demonstrating a statistically significant, positive correlation with local to regional station temperatures. We excluded tree-ring records from lower latitude, elevational treeline sites (e.g., bristlecone pine series from the southwestern USA) that have been thought to demonstrate a possible CO_2 fertilization effect (see below; Glossary; e.g., Wahl and Ammann, 2007).

Table 5.1 Detailed information for regional composite chronologies used to reconstruct extratropical NH temperatures, based on living and subfossil wood material.

Regional Grouping	Code	Full Coverage	No. of series	MSL (years)	Data Type[$]	Period > 10 radii	Chronologies used[#]	RCS strategy*	Reference
NORTH AMERICA									
Seward	**SEW**	978–2002	1196	209	L, S, H	1140–2002	STDPT, RCS	**B**	D'Arrigo et al., 2005
NW North Alaska	**NWNA**	952–2000	294	301	L, S	1297–2000	STD, RCS	**E**	
Yukon	**YUK**	1067–2002	220	265	L, S	1177–2002	STDPT, RCSPT	**E**	
Central Alaska	**CNTA**	1556–1990	51	229	L	1652–1990	STD, RCS	**A**	
Wrangells	**WRA**	1471–1999	159	243	L	1556–1999	STD	**n/a**	Davi et al., 2003
Coastal Alaska	**CSTA**	616–2002	820	242	L, S	713–2002	STD, RCS	**B**	Wiles et al., *in prep*
Central NWT	**CNWT**	1046–2003	569	266	L, S	1288–2003	STD, RCS	**D**	
Southern Alaska	**SA**	1343–2000	143	338	L	1523–1999	STD, RCS	**C**	
Icefields	**ICE**	869–1994	374 (RW) / 153 (MXD)	242 / 232	L, S	918–1994	STDPT, RCSPT	**E**	Luckman and Wilson, 2005
Manitoba	**MAN**	1650–1982	45	267	L	1686–1982	STDPT, RCS	**A**	
Labrador	**LAB**	1459–2001	371	202	L	1570–2001	STD, RCS	**F**	D'Arrigo et al., 2003
Quebec	**QUE**	1404–1991	40	378	L	1504–1991	STD, RCS	**E**	
EURASIA									
Jaemtland	**JAEM**	1106–1978	156	159	L, S	1340–1978	STD, RCS	**E**	Naurzbaev and Vaganov, 1999
Tornetraesk	**TORN**	547–1980	55	309	L, S	747–1980	STD, RCSPT	**A**	Briffa et al., 1992

Polar Urals	**POL**	778–1990	157	162	L, S	944–1990	STD, RCS	Briffa, 2000	Briffa 2000
Taymir	**TAY**	513–1997	236	262	L, S	755–1997	STD, RCS	**B**	Jacoby et al., 2001
Yakutia	**YAK**	1200–1994	179	277	L, S	1342–1994	STD, RCS	**E**	Hughes et al., 1999
Alps	**ALPS**	986–1995	962	126	L, S, H	1350–1995	STDPT, RCS	**E**	Wilson and Topham, 2004; Nicolussi and Schiessling, 2001
Mongolia	**MON**	262–1999	99	341	L, S	913–1999	STDPT, RCSPT	**E**	D'Arrigo et al., 2001

MSL = mean segment length. $: L = Living, S = sub–fossil or snag material and H = historical tree–ring material. #: In select cases, a power transform (PT) was applied to correct for data biases (Cook and Peters, 1997). This bias was assessed by correlation and residual analysis against both local and large–scale temperature series. *: Due to differing populations in the tree–ring (TR) data, the datasets were often grouped into 'common' populations (Esper et al., 2003). No one strategy is appropriate for all datasets and careful evaluation of each composite dataset was made. Strategy codes = **A:** One regional curve; **B:** Multiple regional curves, related to growth trend type (i.e., those series traditionally detrended using negative exponential or linear regression function); **C:** Multiple regional curves, related to growth level differences (i.e., data were divided equally into groups of low and high growth rates); **D:** Mixture of B and C; **E:** Multiple regional curves, data divided 'horizontally' into separate living and subfossil (and historical) sub groups; **F:** Separate regional curves for each constituent chronology. The relevant original references for each dataset are listed. Note for the Alps, different datasets were utilized for the standard power transform (STDPT) and Regional Curve Standardization (RCS) chronology versions. The STDPT is a highly replicated expanded dataset of that detailed in Wilson and Topham (2004). RCS was not possible, however, with these data due to significant differences in mean RW between the historical and living datasets which resulted in a highly biased RCS chronology. Instead, for the Alps RCS chronology, the Nicolussi and Schiessling (2001) long pine RCS chronology was used. No STD version of this chronology exists. Adapted from D'Arrigo et al. (2006).

Table 5.2 Coldest and warmest decades (anomaly values in parentheses, degrees C) calculated over 1000–1979 for each NH temperature reconstruction after they have been scaled (w.r.t. 1856–1978) to NH land only (20–90°N) mean annual temperatures. The difference between these two values is defined as the amplitude. Adapted from D'Arrigo et al. (2006).

	Coldest	Warmest	Amplitude
This Study STD	1813–1822 (−0.74)	1938–1947 (0.20)	0.94
This Study RCS	1600–1609 (−0.97)	1937–1946 (0.17)	1.14
Esper, 2002	1345–1354 (−1.18)	1950–1959 (0.15)	1.34
Briffa, 2000	1813–1822 (−0.80)	1951–1960 (0.10)	0.90
Mann, 1999	1458–1467 (−0.68)	1957–1966 (0.10)	0.79
Jones, 1998	1693–1702 (−0.77)	1929–1938 (0.06)	0.83
Moberg, 2005	1576–1585 (−1.33)	1104–1113 (0.23)	1.56

This work was presented before the National Research Council in the year of its publication, with the results appearing in a subsequent set of recommendations and follow-up reports (NRC, 2006, and a report by the Pew Center for Climate Change, unpublished). These results also appeared in the subsequent IPCC report (Jansen et al., 2007) as well as other publications of potential interest to the general public and media outlets (e.g., featured in *New Scientist* March 18, 2006 and online in 2009: http://www.newscientist.com/article/dn11646-climate-myths-the-hockey-stick-graph-has-been-proven-wrong.html).

Figure 5.1 shows the individual 66 tree-ring sites that were used in this compilation, and Table 5.1 lists the regional chronologies that were generated based on individual site data that were then averaged to yield the extratropical NH composite. Figures 5.2–5.4 illustrate the temperature reconstructions and related analyses and comparisons.

These raw measurements and tree-ring chronologies have been used by researchers in the fields of dendrochronology and climate modeling, helping to form the baseline for reconstructions and modeling of climate. Progressing beyond the initial development of relatively simple reconstructions using linear regression techniques, several studies have incorporated such data into analyses using relatively complex methods, such as climate field reconstructions (CFR; Glossary) of past spatial patterns and climate fields, data-model comparisons of past large-scale temperature and climate variability, comparison of regression and scaling methods, and assessment of non-linear responses to environmental change (Christiansen et al., 2009; Christiansen and Lunjqvist, 2012; Esper et al., 2005; Mann et al., 2005; Mann et al., 2007a; Mann et al., 2007b; Rutherford et al., 2005; Smerdon and Kaplan, 2007; Smerdon et al., 2010; Von Storch et al., 2004; Von Storch et al., 2009).

Notably, a newly integrated compilation of the various large-scale proxy temperature reconstructions, spanning the past one to two millennia, has been

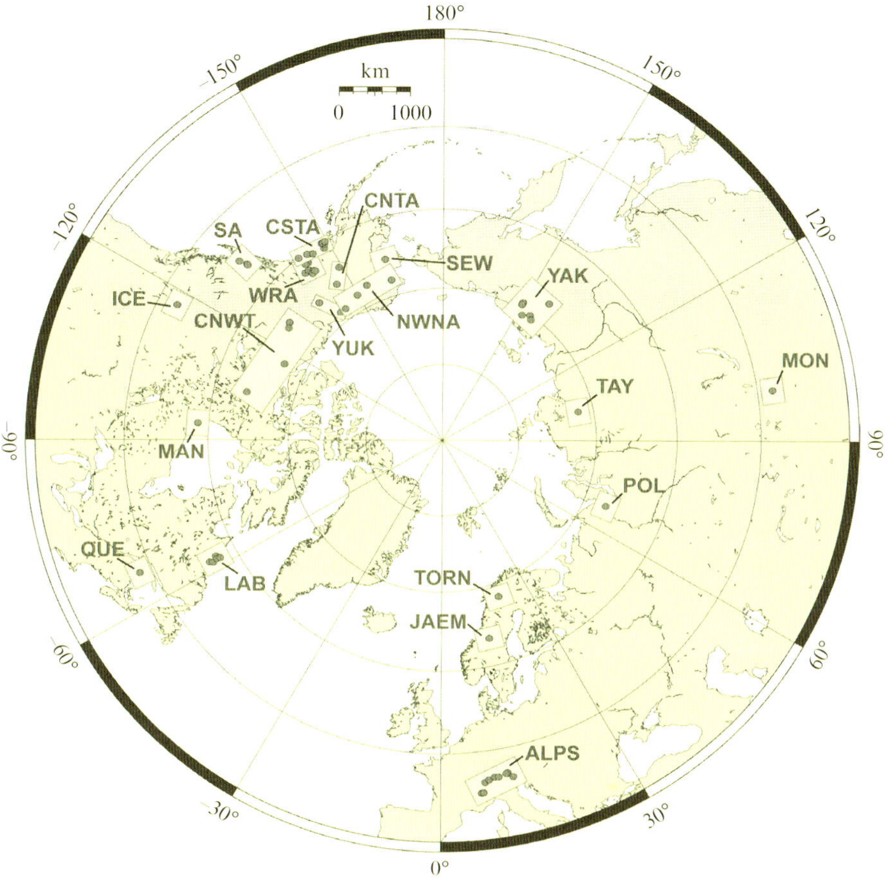

Figure 5.1 Location map of individual tree-ring sites (*red*) and regional composites (*yellow boxes*) used to reconstruct NH temperatures over the past millennium. See Table 5.1 for more details. Note that there are many tree-ring records generated for North America and Eurasia by the dendroclimatic community that are not included in this figure. Adapted from D'Arrigo et al., 2006. For color detail, please see color plate section.

recalibrated into a single product. It is now available online at the NOAA National Climatic Data Center, and is known as the Paleoclimatology Reconstruction Network or PCN (Wahl et al., 2010; http://www.ncdc.noaa.gov/paleo/pubs/pcn/pcn.html). This compilation includes metadata and recalibration of data to a common standard, spanning local to global scales. Wahl and colleagues (2010) also estimate the amplitude and rate of change of the globe's response to anthropogenic greenhouse gases, and provide an analysis of the inherent uncertainties in such estimates and their spectral properties.

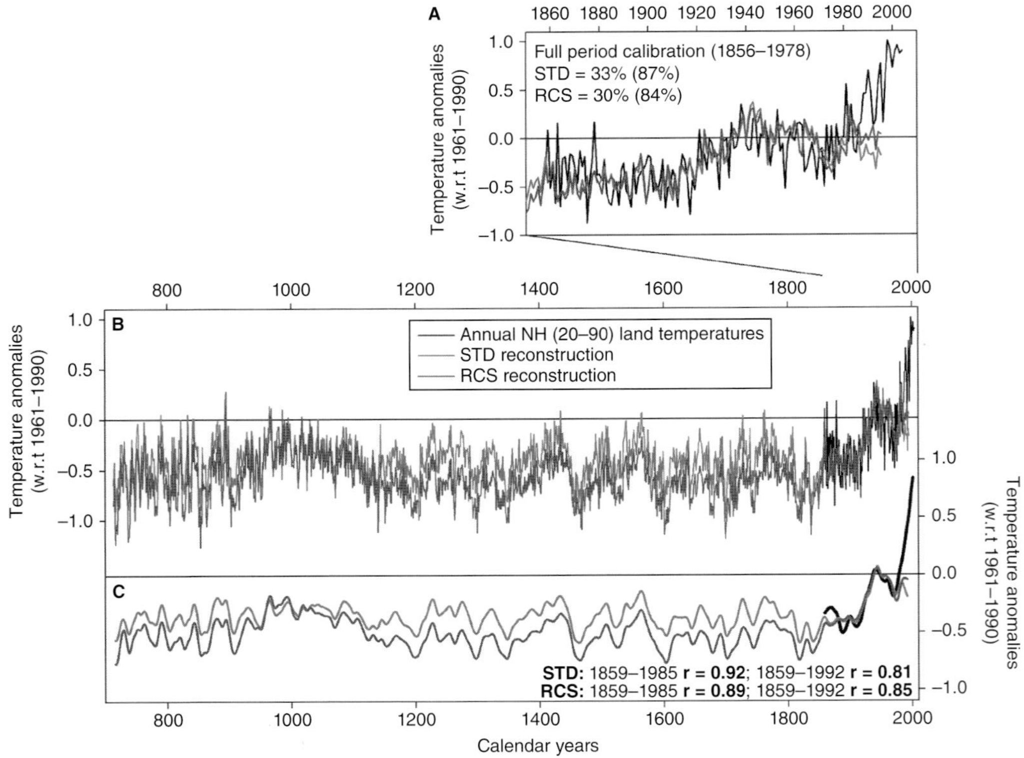

Figure 5.2 Comparison of STD and RCS NH reconstructions with mean annual land (20–90°N) temperatures. **A**. Calibration period values denote the variance explained from the calibration. Values in parentheses are for 20-year smoothed filtered versions. **B**. Full length time series. Although regression-based techniques can systematically underestimate low-frequency trends (Von Storch et al., 2004), we have partially overcome this problem by scaling (w.r.t 1856–1978) the calibrated reconstructed time-series to the instrumental record (Esper et al., 2005). **C**. as in **B** but using smoothed (20-year) time series. The values show correlations between the smoothed reconstructions and the instrumental records for the extended periods 1859–1985 and 1859–1992. To reduce potential end effect biases of the smoothed series, three years were truncated from the ends of the time series before correlation analysis. Adapted from D'Arrigo et al., 2006. For color detail, please see color plate section.

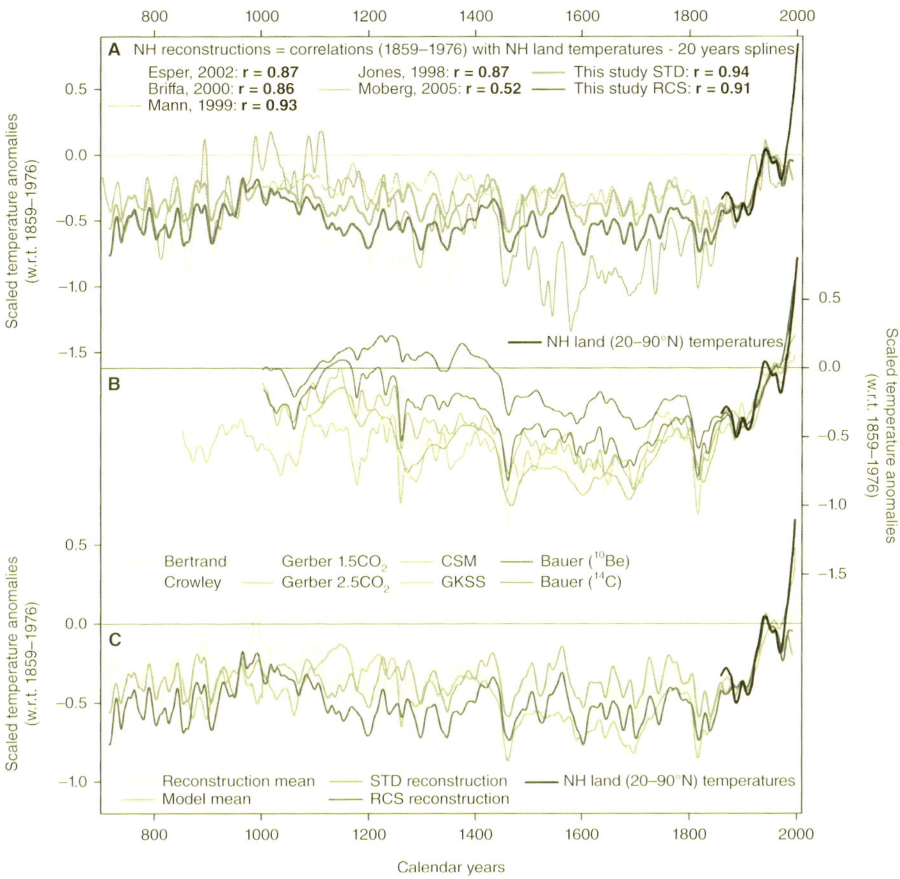

Figure 5.3 A. Comparison of STD and RCS NH reconstructions with previous reconstructions. **B**. Model based estimates of NH temperatures for the last millennium (Jones and Mann, 2004). **C**. Comparison of mean series of the previously developed reconstructions and models with the STD and RCS series. The reconstruction and model time-series were normalized to the common period and averaged. All smoothed series in this figure were scaled to the smoothed instrumental NH temperature series over the period 1859–1976. Adapted from D'Arrigo et al., 2006. For color detail, please see color plate section.

5.3. Reconstructed NH Temperature Trends

Notable time periods seen to varying degrees in these large-scale temperature reconstructions include the broadly defined intervals known as the so-called Medieval Climate Anomaly or MCA (Glossary), early in the past millennium (mid tenth to thirteenth centuries: Diaz et al., 2010), the Little Ice Age (LIA,

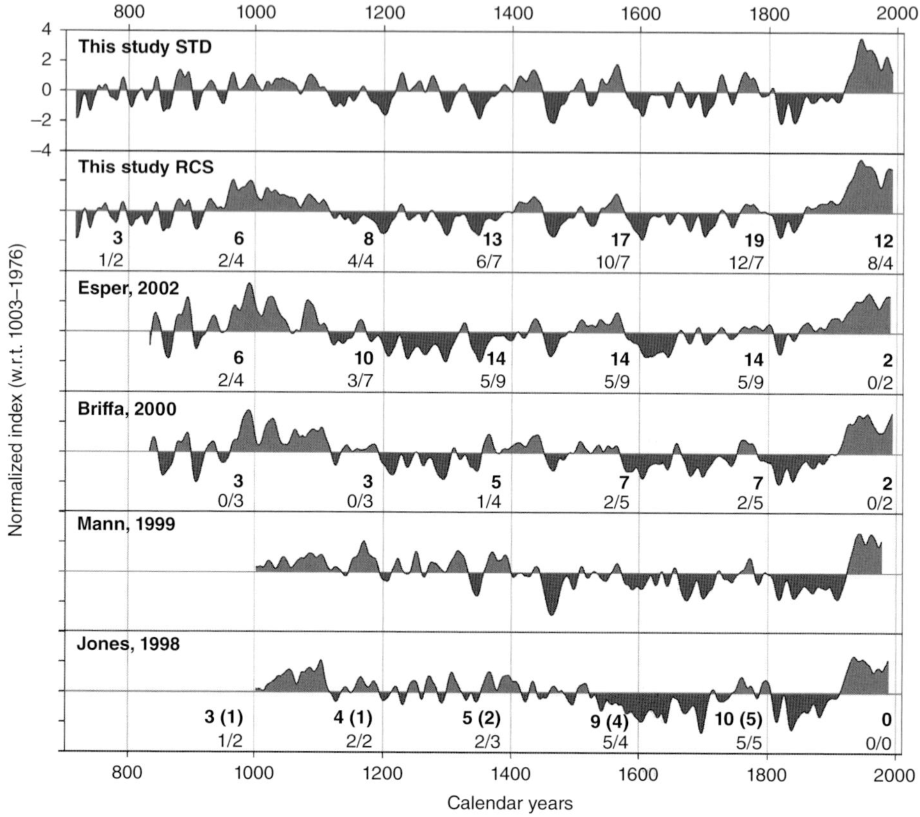

Figure 5.4 Smoothed (20-year spline) NH temperature reconstructions. Y axis values in degrees C. The time series have been normalized to their common period. The values denote replication (number of chronologies/regional composite series) at the beginning of each 200-year window. The most recent values are for the year 1995 to highlight the updated data-sets used for this study. The upper bold values show the full number of series. Value in parentheses denote those series used that were not tree-ring data. The lower values detail the number of records from North America/Eurasia, respectively. We have not included equivalent information for the Mann et al. (1999) study because it used principal component regression, which resulted in an unequal weighting of the initial records in the final reconstructions. Mann et al. (1999) used 112 records in the most replicated period. This decreased to 7 records by 1000 AD (6 of which were tree-ring data). The numbers for 1500 and 1000 include principal components derived from 21 western and 6 southern U.S. tree-ring sites that are counted as 2 regional records. Adapted from D'Arrigo et al., 2006. For color detail, please see color plate section.

~1450 to 1850, Grove, 1988), and the transition between the MCA and the LIA (Trouet et al., 2009; Frank et al., 2010). Although these terms can be descriptive and are still in common usage, some researchers have called for their abandonment or caution in their usage, as the particular warm or cold anomalies of interest can vary considerably in space and in time (Jones et al., 1998; Jones and Mann, 2004; D'Arrigo et al., 2006).

Virtually all of the published large-scale temperature reconstruction studies, despite using different methodologies and having only partially overlapping data sets, have reached broad consensus that recent warming in the NH appears to have been unprecedented over the past millennium and that this warming is most likely a result of the anthropogenic release of greenhouse gases into the atmosphere (Figures 5.2–5.4; IPCC, 2007; see Frank et al., 2010 for a review). The unusual nature of reconstructed late twentieth century temperatures is a typically robust feature even when individual series are systematically excluded, and the hemispheric reconstructions generated thus far largely fall within each other's respective uncertainty limits (IPCC, 2007). The overall trends of a warm MCA early in the past millennium, transitioning to the colder LIA, followed by a recovery and recent warming are generally observed in all of these reconstructions whether they be based solely on tree-ring data (Esper et al., 2002; D'Arrigo et al., 2006) or are multi-proxy in approach (Moberg et al., 2005; Mann et al., 2008; Ljungqvist, 2010; Christiansen and Ljungqvist, 2012).

There are nevertheless uncertainties and disagreements in published efforts to depict past large-scale temperature variability, especially related to amplitude change, due to differences in site coverage, proxy type, methodology, and other factors. A considerable range in reconstructed amplitudes is observed among the existing NH temperature reconstructions, particularly with regard to the transition from the MCA into the LIA (Table 5.2, Figures 5.2–5.4, D'Arrigo et al., 2006; IPCC, 2007). This is largely due to the fact that data coverage is typically low during this interval in virtually all existing reconstructions.

5.4. Standardization of NH Tree-Ring Temperature Reconstructions

D'Arrigo and colleagues (2006) quantified differences between more traditional (STD) and Regional Curve Standardization (RCS) methodologies. It was concluded that RCS was generally superior for retention of low-frequency trends in this data set, based on our own visual observations and comparisons with instrumental data and model simulations of temperature. This reconstruction (Figures 5.2–5.4) indicates clear MCA (warm), LIA (cool), and recent (warm) episodes. Direct interpretation of the RCS reconstruction suggests that MCA temperatures were nearly 0.7 °C cooler than the late twentieth century, with an amplitude difference of 1.14 °C from the coldest (1600–1609) to warmest

(1937–1946) decades (Table 5.2). Note that the more recent decades are warmer but are not directly included in this estimate, because the common period of the tree-ring data ends sooner, and because there are some divergence issues in the most recent period of the reconstruction (Wilson et al., 2007b). The recent warming appears to be quite substantial relative to natural fluctuations of the past millennium. However, as described by D'Arrigo and colleagues (2006), the spatially-heterogeneous nature of the MCA, and its different timing within the different regional composites, results in a 'flattening out' of the large-scale estimates for this period relative to twentieth century warming, with the latter expressing a more homogenous global 'fingerprint.' The spatial coverage and sample depth of the tree-ring records also decreases during the MCA period, limiting robust comparison and interpretation at this time.

In another large-scale temperature reconstruction for the NH, Esper and colleagues (2002, updated in Frank et al., 2007) also used the Regional Curve Standardization (RCS) method to retain multicentennial temperature variability. Esper and colleagues (2002) averaged ring-width chronologies from North American and Eurasia, which supported the large-scale occurrence of the MCA over the northern land-only extratropics. Although persistently above-average temperatures in the AD 960–1050 interval suggested an MCA, declining site availability, and low within-chronology replication prior to AD 1200 were found to weaken this interpretation considerably (Cook et al., 2004).

We have concluded that currently available paleoclimatic reconstructions may be inadequate to make specific inferences, at hemispheric scales, about MCA warmth relative to the present anthropogenic period. More studies and data are needed to better understand the lower-frequency variability in such reconstructions and the climate signatures reflected in these data.

The scientific process of generating dendroclimatic reconstructions based on tree-ring width, density, isotopic, and other parameters is still evolving, and is becoming more complex. One reason is that data coverage has now improved to the point that large-scale networks of tree-ring data are available for use in spatial climate field reconstructions, providing opportunities to model atmosphere-ocean-land climate variability and dynamics (e.g., Rutherford et al., 2005; Mann et al., 2008; Mann et al., 2009). Another concern is the ongoing debate on how best to standardize tree-ring series to optimize retention of low-frequency trends (e.g., Briffa and Melvin, 2010). Another aspect of this continued evolution has been the generation of forward modeling techniques (Glossary) to better understand the mechanisms of tree growth response to climate and environmental forcing (Vaganov et al., 2006; Anchukaitis et al., 2006; Evans et al., 2006). Forward modeling employs a mechanistic model of tree-ring formation to simulate patterns of climate-tree growth relationships and test for biological or ecological mechanisms that may cause divergence or other anomalous or nonlinear patterns to occur. These models can simulate tree growth over a range of climatic regimes in order to aid in understanding the critical physiological processes of forest growth response to climate and other environmental influences (Anchukaitis et al., 2006).

Studies are also under way to integrate climate proxies and modeling using data-model comparison and data assimilation techniques (e.g., Goose, 2010; Graham et al., 2010). These latter efforts include evaluation of real and synthetic model-generated proxies, comparison of proxies to transient climate model runs, optimal ensemble simulations, Bayesian hierarchical models (Tingley et al., 2010a; Tingley et al., 2010b), and time-dependent methodologies such as Kalman filter ensembles (Visser and Molenaar, 1990; Visser et al., 2010).

Model comparisons, based on General Circulation Model (GCM) simulations, show reasonable coherence over the last 600 years with the NH temperature reconstruction shown in D'Arrigo and colleagues (2006, Figure 5.3; Mann et al., 2012). Proxy reconstructions, however, generally show an earlier peak in MCA warmth compared to models (see Jones and Mann, 2004; D'Arrigo et al., 2006), possibly reflecting that this was a spatially-complex, highly-variable period, and that not enough proxy records yet exist for this time. It is also possible that the models themselves may be biased in various ways. For example, although they incorporate external (solar, volcanic, anthropogenic) forcings, they may not capture internal atmosphere-ocean dynamics that well.

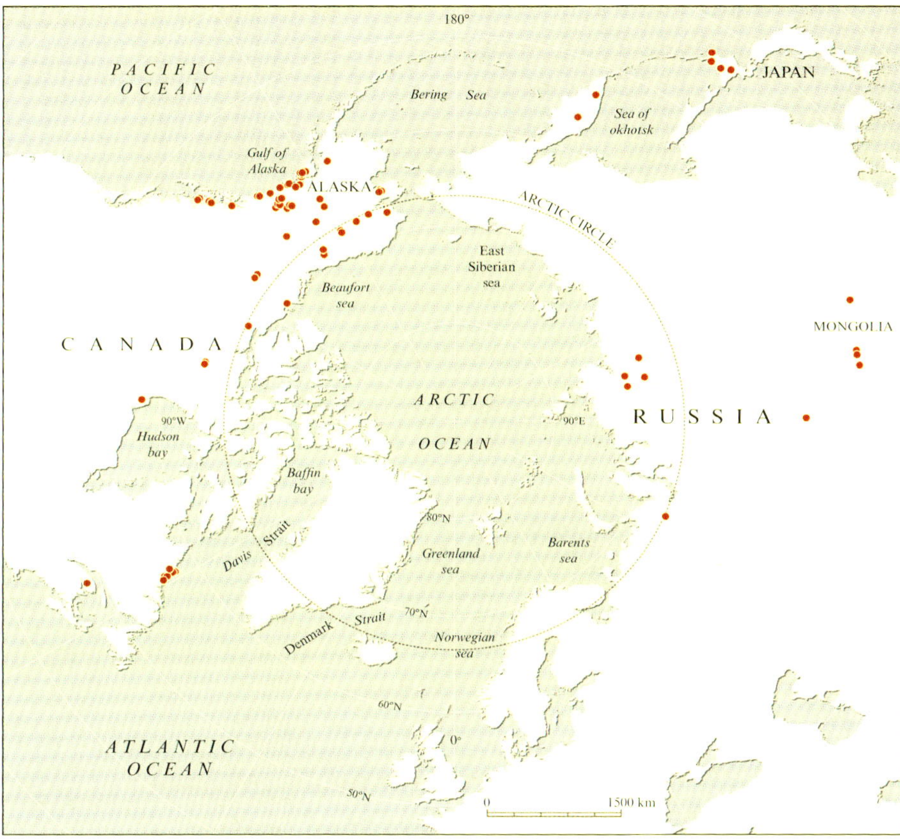

Figure 1.1 Locations of tree-ring chronologies sampled from northern forest sites over the past few decades by TRL-LDEO scientists and colleagues. Many more have been sampled by other scientists for these areas.

Dendroclimatic Studies: Tree Growth and Climate Change in Northern Forests, Monograph Number 67. Rosanne D'Arrigo, Nicole Davi, Gordon Jacoby, Rob Wilson, and Greg Wiles.

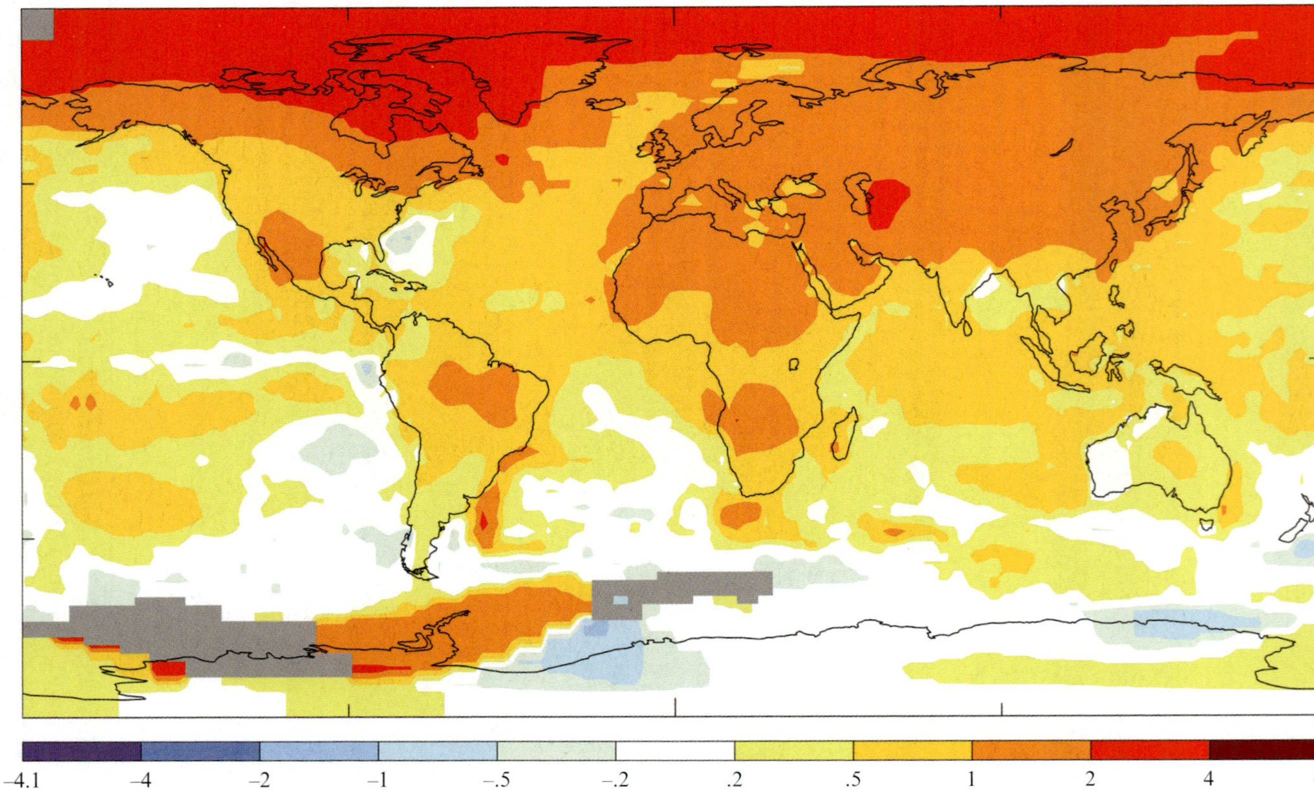

Figure 1.2 Map showing evidence from annual (Jan–Dec) temperature data of large-scale warming and polar amplification in recent decades (temperatures averaged over 1975–2010 in degrees Celsius). Adapted from Goddard Institute for Space Studies (GISS) (http://data.giss.nasa.gov/gistemp/maps/; Hansen et al., 2010).

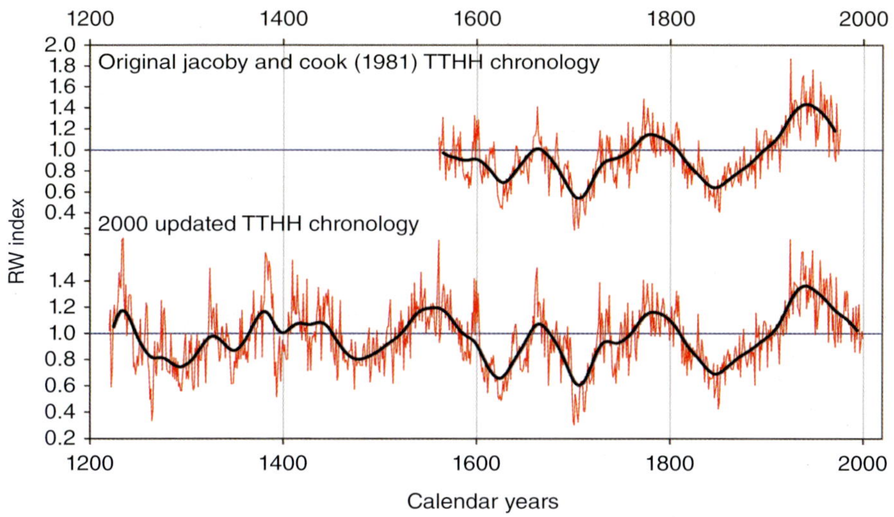

Figure 3.1 Original (*top*) and updated Twisted Tree-Heartrot Hill (TTHH) tree-ring width chronologies from elevational treeline, Yukon Territory, Canada. Units are in dimensionless indices. Adapted from Jacoby and Cook (1981).

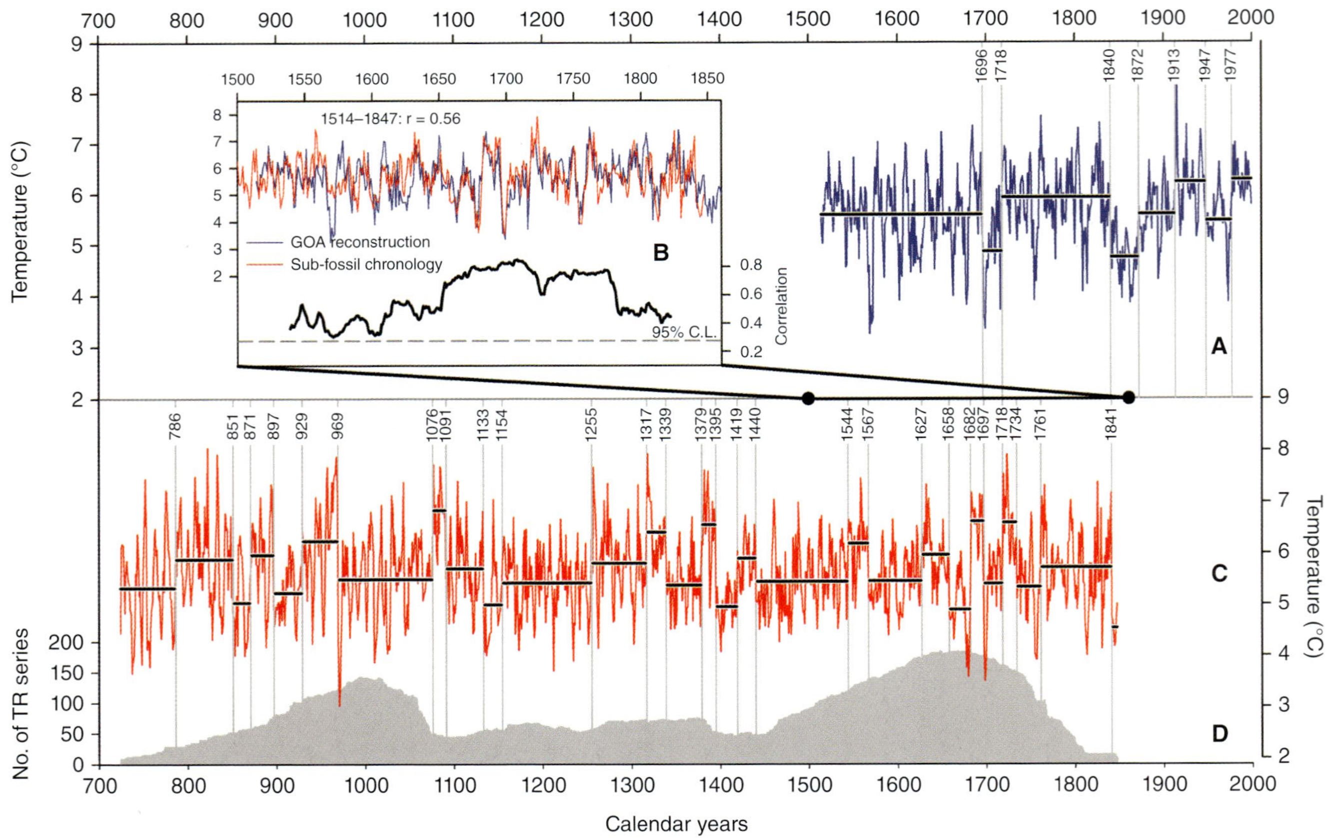

Figure 3.2 Comparison of Gulf of Alaska (GOA) living and subfossil tree-ring width chronologies. The subfossil data have been scaled to the reconstruction over the period of overlap (1514–1847). **A**. The GOA reconstruction time series. **B**. Period of overlap with sliding 31-year correlations. **C**. The scaled subfossil chronology. The vertical lines (with dates) in **A** and **C** show significant (95% confidence level) regime shifts in the time series interpreted as representing decadal variability in North Pacific climate. **D**. Series replication in the subfossil chronology. Adapted from Wilson et al. (2007a).

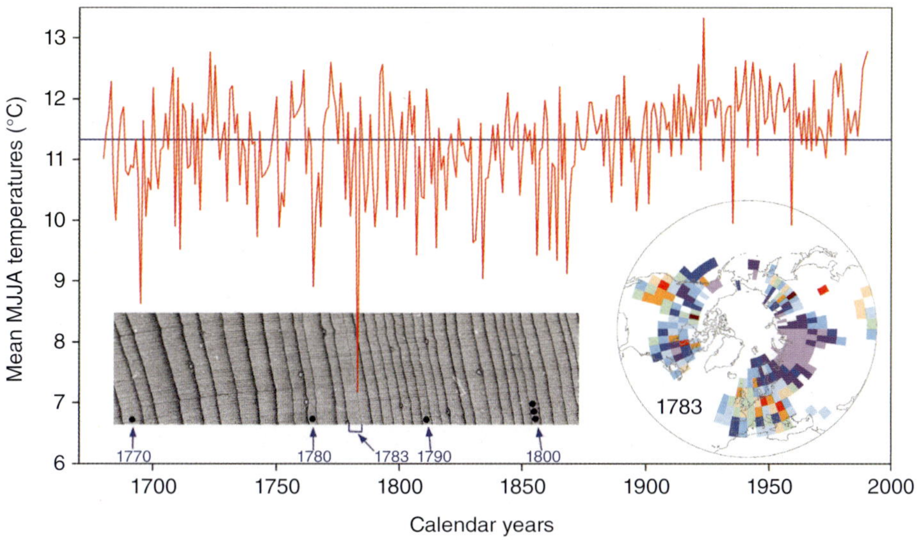

Figure 3.3 Tree-ring reconstruction of May–August temperatures for Alaska based on maximum latewood density data. Note anomalously low reconstructed temperatures during 1783, the year of the major Laki volcanic eruption in Iceland. Adapted from Jacoby et al. (1999).

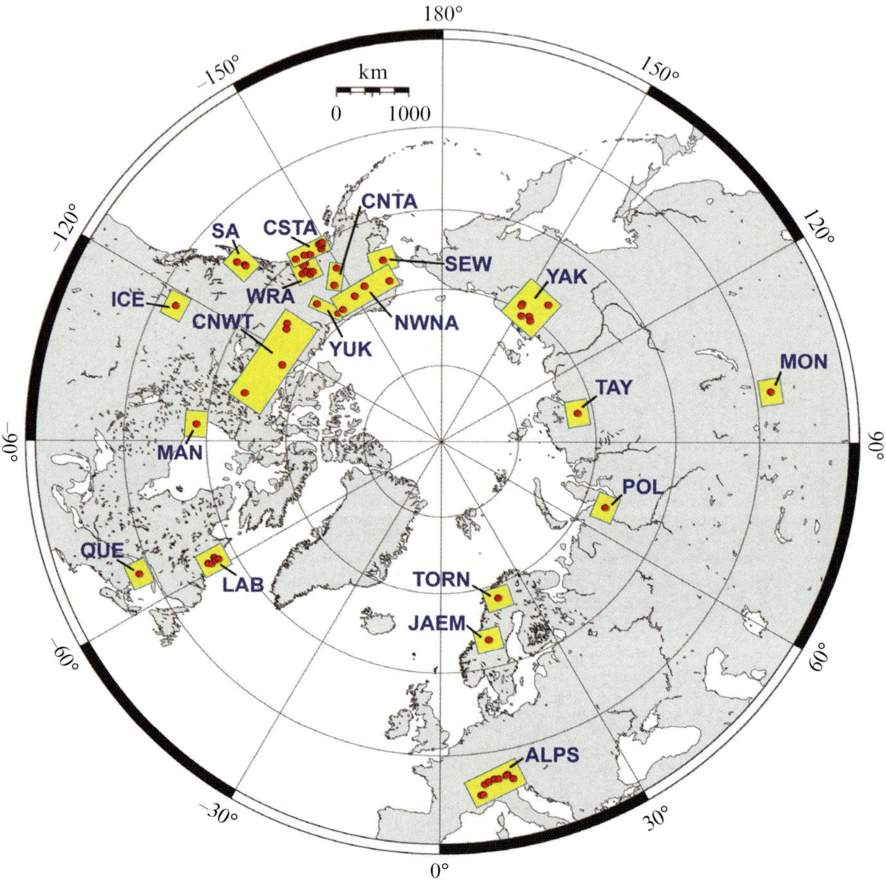

Figure 5.1 Location map of individual tree-ring sites (*red*) and regional composites (*yellow boxes*) used to reconstruct Northern Hemisphere temperatures over the past millennium. See Table 1 for more details. Note that there are many tree-ring records generated for North America and Eurasia by the dendroclimatic community that are not included in this figure. Adapted from D'Arrigo et al., 2006.

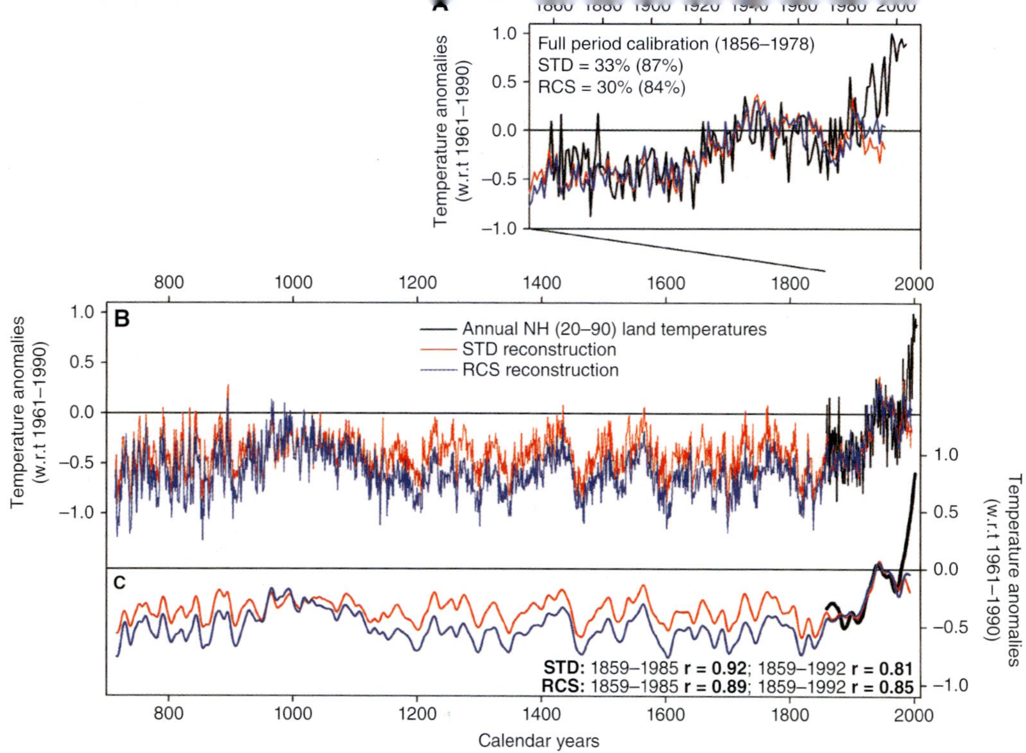

Figure 5.2 Comparison of STD and RCS NH reconstructions with mean annual land (20–90°N) temperatures. **A**. Calibration period values denote the variance explained from the calibration. Values in parentheses are for 20-year smoothed filtered versions. **B**. Full length time series. Although regression-based techniques can systematically underestimate low-frequency trends (Von Storch et al., 2004), we have partially overcome this problem by scaling (w.r.t 1856–1978) the calibrated reconstructed time-series to the instrumental record (Esper et al., 2005). **C**. as in **B** but using smoothed (20-year) time series. The values show correlations between the smoothed reconstructions and the instrumental records for the extended periods 1859–1985 and 1859–1992. To reduce potential end effect biases of the smoothed series, three years were truncated from the ends of the time series before correlation analysis. Adapted from D'Arrigo et al., 2006.

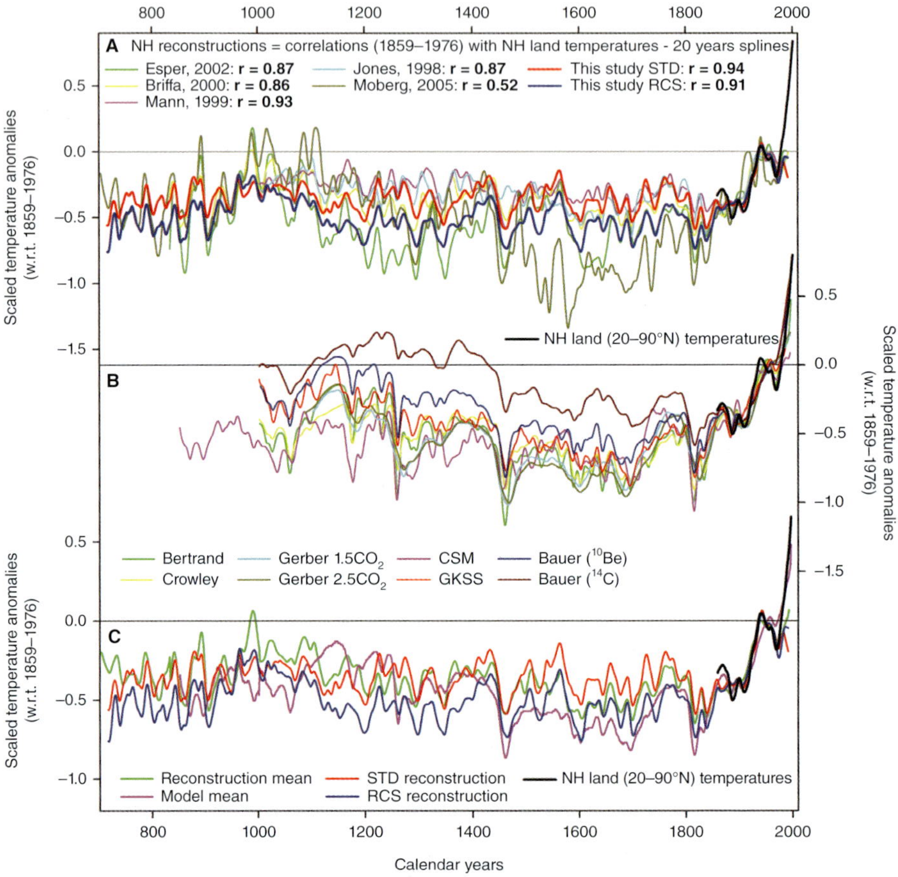

Figure 5.3 A. Comparison of STD and RCS NH reconstructions with previous reconstructions. **B**. Model based estimates of NH temperatures for the last millennium (Jones and Mann, 2004). **C**. Comparison of mean series of the previously developed reconstructions and models with the STD and RCS series. The reconstruction and model time-series were normalized to the common period and averaged. All smoothed series in this figure were scaled to the smoothed instrumental NH temperature series over the period 1859–1976. Adapted from D'Arrigo et al., 2006.

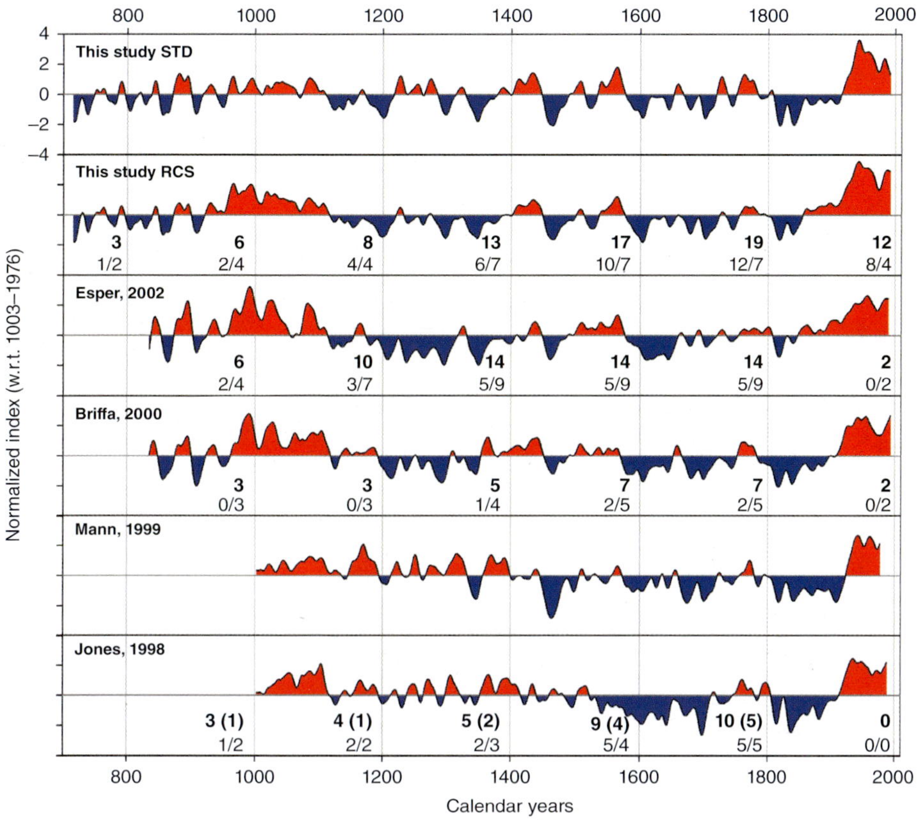

Figure 5.4 Smoothed (20-year spline) NH temperature reconstructions. Y axis values in degrees C. The time series have been normalized to their common period. The values denote replication (number of chronologies/regional composite series) at the beginning of each 200-year window. The most recent values are for the year 1995 to highlight the updated data-sets used for this study. The upper bold values show the full number of series. Value in parentheses denote those series used that were not tree-ring data. The lower values detail the number of records from North America/Eurasia, respectively. We have not included equivalent information for the Mann et al. (1999) study because it used principal component regression, which resulted in an unequal weighting of the initial records in the final reconstructions. Mann et al. (1999) used 112 records in the most replicated period. This decreased to 7 records by 1000 AD (6 of which were tree-ring data). The numbers for 1500 and 1000 include principal components derived from 21 western and 6 southern U.S. tree-ring sites that are counted as 2 regional records. Adapted from D'Arrigo et al., 2006.

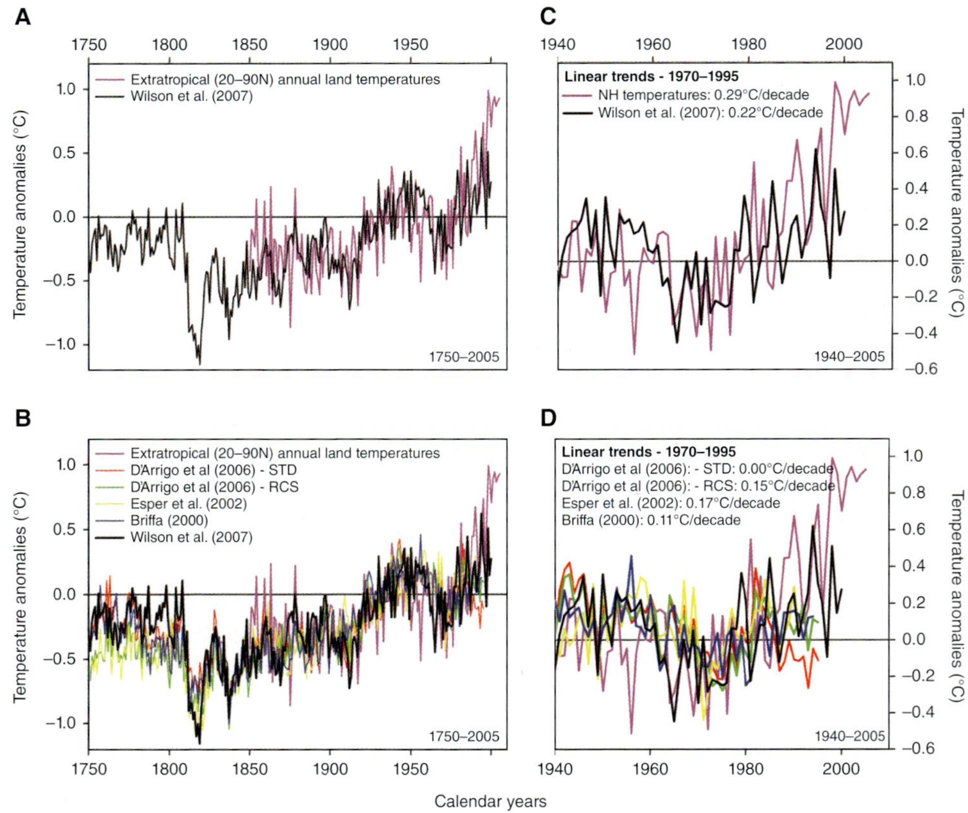

Figure 6.1 Extratropical instrumental and reconstructed (20–90°N) NH annual land temperatures. **(A)** Wilson et al., 2007b, and instrumental data, 1750–2005. **(B)** Wilson et al., 2007b, instrumental data and previous TR based extratropical NH reconstructions, 1750–2005. **(C)** For 1940–2005. Linear trends are calculated for the 1970–1995 period. **(D)** For 1940–2005. Note that although the linear trends are calculated over the 1970–1995 period, the ECS2002 and BRF2000 series only extend to 1992 and 1994, respectively. All of these comparisons illustrate a disparity or divergence between the tree-ring based proxies and instrumental temperatures, with the trees underestimating trends over recent decades.

A

B

Figure 7.1 Photographs of northern tree-ring field investigations and the authors. **A**. The very first tree (Siberian pine) cored in Mongolia (1475–1999) in 1995. The data from this tree and others at this site resulted in a publication on large-scale warming in Mongolia (Jacoby et al., 1996, *Science*). (Photo Credit: R. D'Arrigo) **B**. A cross section from a wood sample found in the Tarvagatay Mountains of Mongolia. The inner pith dates to 262 CE. (Photo Credit: N. Pederson)

C

D

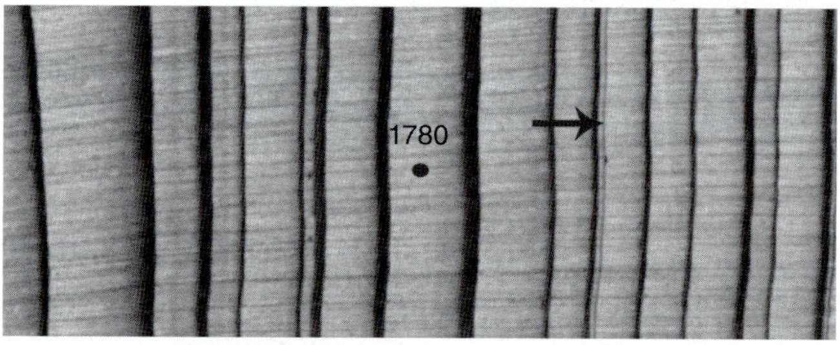

Figure 7.1 **(*Cont'd*)** **C**. A Piper Cub plane getting ready to land in a makeshift runway of willow scrub in Wrangell St. Elias National Park and Preserve, Alaska. Other than trekking, a Piper Cub is the only way into the park. (Photo Credit: N. Davi) **D**. A tree core from Alaska showing a VERY NARROW, light latewood band at the year 1783. The Laki volcanic eruption occurred in Iceland during that year (in June) and cooled global temperatures, causing this tree to stop growing abruptly in the fall of that year (Jacoby et al., 1999). (Photo Credit: G. Jacoby)

E

F

Figure 7.1(*Cont'd*) E. Dr. Jacoby stands on top of ancient White River Ash, a layer of ash from Mt. Churchill that blanketed more than 340,000 sq km of eastern Alaska and NW Canada in 800AD. Tree stumps buried by the event were exposed as the bank eroded. These samples, along with living white spruce, were used to extend our understanding of climate back by nearly two thousand years. (Photo Credit: N. Davi) **F**. 400-year-old Siberian larch trees growing in the Mongolian steppe. These trees are growing in very arid conditions and are sensitive to changes in moisture. (Photo Credit: B. Buckley)

G

H

Figure 7.1(*Cont'd*) G. Davi in a white spruce forest overlooking Mendenhall Glacier in Juneau, AK. These trees were used to learn about climate change and glacier dynamics over the past 400–500 years. (Photo Credit: G. Jacoby) **H**. Drs. Gordon Jacoby and Rosanne D'Arrigo sampling native trees in Alaska in the early 1980s. (Photo Credit: Unknown)

I

Figure 7.1(*Cont'd*) I. Dr. Greg Wiles sampling trees near tree line in Glacier Bay National Park and Preserve, Alaska. (Photo Credit: Unknown)

J

Figure 7.1(*Cont'd*) J. Rob Wilson beside relict root stocks (preserved in peat) of ancient Scots pine trees in Glen Affric (Scottish highlands) at about 550 m above current treeline. As yet, this location is not radiocarbon dated, but a similar site in the Cairngorms (90 km to east) has been dated to ~6000 years BP. (Photo Credit: A. Wilson)

6

6. TREE GROWTH ISSUES IN THE ANTHROPOGENIC ERA: CO_2 FERTILIZATION AND THE "DIVERGENCE PROBLEM"

In addition to our paleoclimatic studies, TRL-LDEO scientists, along with other investigators, have analyzed the response of northern forest growth, physiology, and ecology to climatic and environmental change. When developing temperature reconstructions from tree rings, for example, routine examination and testing of residual estimates from regression models have been used to reveal whether unexplained trends, possibly due to CO_2 fertilization (Jacoby and D'Arrigo, 1997; LaMarche et al., 1984) or divergence effects (D'Arrigo et al., 2008) may exist.

6.1. CO_2 Fertilization

The hypothesis that increasing atmospheric CO_2 levels since the middle nineteenth century may have directly contributed to increased radial growth of bristlecone and limber pine trees at elevational treeline in the southwestern United States that is unexplained by climate was first put forward by LaMarche and colleagues (1984). It was later suggested that only those trees that featured a so-called "strip-bark" morphology would be expected to show such unexplained growth enhancement (Graybill and Idso, 1993; Ababneh, 2006). This topic remains controversial, in part because long instrumental climate station records that can be used for direct comparison to the tree growth variations are relatively scarce in remote elevational or latitudinal treeline regions where many temperature sensitive tree-ring chronologies have been developed. It is also difficult to separate out the climate and CO_2 influences on tree growth, as both show increasing trends over the last 150 years. Further, field experiments indicating fertilization-type effects are not necessarily representative of the response of the old growth boreal forests (e.g., Hickler et al., 2008). Thus far, we have found no such evidence for CO_2 fertilization at the northern sites we have studied, perhaps

Dendroclimatic Studies: Tree Growth and Climate Change in Northern Forests,
Monograph Number 67. Rosanne D'Arrigo, Nicole Davi, Gordon Jacoby,
Rob Wilson, and Greg Wiles.

due to the substantial nutrient limitations that can occur at far northern latitudes that preclude the trees' ability to respond to increasing CO_2 concentrations (Jacoby and D'Arrigo, 1997).

6.2. The Divergence Problem

Over the recent decades of rapid warming and significant environmental change, we have observed evidence for a decoupling between tree growth and temperature trends at many northern sites (e.g., Figure 6.1).

Initially described by Jacoby and D'Arrigo (1995), divergence has potentially important implications for the interpretation of paleoclimatic reconstructions and their inferred temperature sensitivity, although its scope and severity still remain uncertain. This phenomenon is generally described for the low-frequency time scale, and may or may not be present in year-to-year comparisons with temperature (Esper and Frank, 2009).

Divergence effects appear to be concentrated in the northern boreal forests (Briffa et al., 1998a; Briffa et al., 1998b; Vaganov et al., 1999; D'Arrigo et al., 2008). A key theory is that large-scale warming is causing temperature-induced drought stress, even at previously temperature–limited northern sites, although other factors may also play a role (see discussion below). Divergence-type effects seem particularly prevalent for central to northern Alaska and vicinity, a region that as noted is warming faster than most of the globe (e.g., Jacoby and D'Arrigo, 1995; Barber et al., 2000; Juday, 2011).

An apparent decrease in recent temperature sensitivity is evident in virtually all tree-ring based NH reconstructions, with a concomitant decrease/flattening in post-1970 trends (Figure 6.1; see also Wilson et al., 2007b). There is also a general weakening in recent decades in calibration results and verification statistics typically employed in dendroclimatology, such as in the level of explained common variance when calibrating with climate data, and when verifying using the Reduction of Error (RE) and the Coefficient of Efficiency (CE) (Cook and Kairiukstis, 1990). Comparison between northern treeline reconstructions of temperature based on tree rings and those for middle northern latitudes suggests that the divergence is restricted to the most recent decades, when these two time series begin to diverge (Cook et al., 2004).

Despite the previous observations, there are many northern sites where there is no apparent divergence (see also Wilson et al., 2007b; Esper et al. 2010), for example, for the MXD record from the Firth River area of northeastern Alaska, which shows a positive response to temperature over the past century (Anchukaitis et al., 2012b). A millennial-length tree-ring reconstruction for the Gulf of Alaska region (Wiles et al., in review) has revealed an absence of divergence with temperature for trees growing at moderate elevations, whereas those

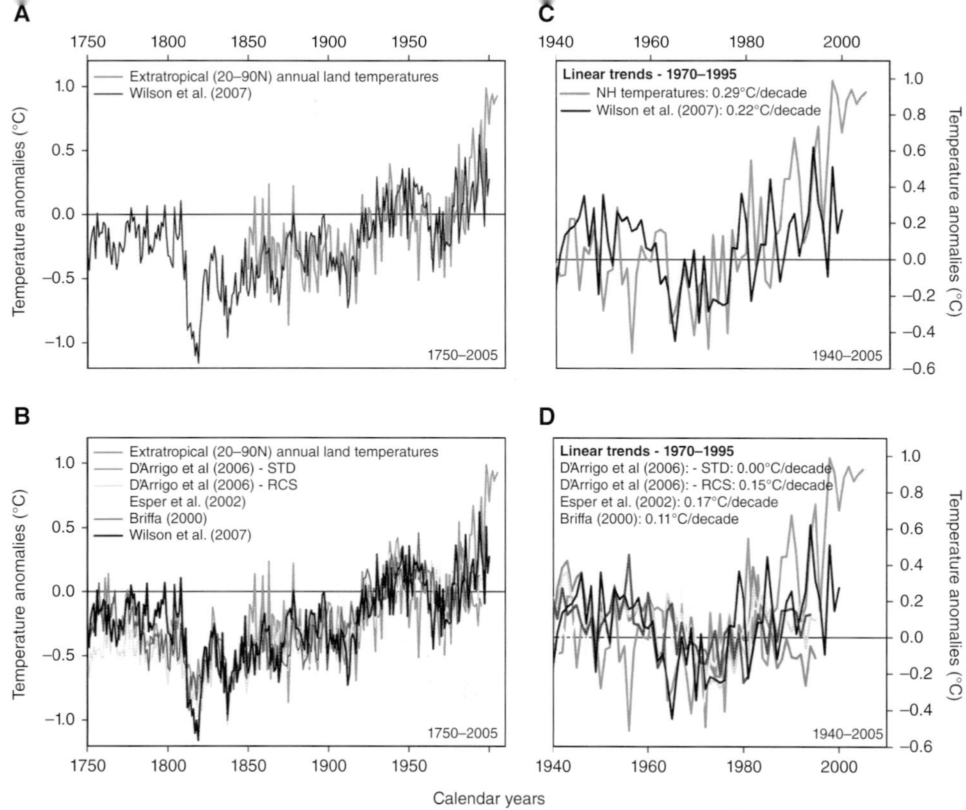

Figure 6.1 Extratropical instrumental and reconstructed (20–90°N) NH annual land temperatures. **A.** Wilson et al., 2007b, and instrumental data, 1750–2005. **B.** Wilson et al., 2007b, instrumental data and previous TR based extratropical NH reconstructions, 1750–2005. **C.** For 1940–2005. Linear trends are calculated for the 1970–1995 period. **D.** For 1940–2005. Note that although the linear trends are calculated over the 1970–1995 period, the ECS2002 and BRF2000 series only extend to 1992 and 1994, respectively. All of these comparisons illustrate a disparity or divergence between the tree-ring based proxies and instrumental temperatures, with the trees underestimating trends over recent decades. For color detail, please see color plate section.

at lower and higher elevations do show this effect. This is one potential strategy for avoiding this phenomenon so long as the sub-fossil tree-ring data are also sampled from similar elevations.

Such series can be combined to derive near "divergence-free" large-scale tree-ring based reconstructions of large-scale temperature (Figure 6.1) (Wilson et al., 2007b). The Wilson and colleagues (2007b) reconstruction was derived using an independent network of temperature-sensitive tree-ring data that had not been utilized in previous large-scale composite studies. Although this paper provided strong independent verification of previous hemispheric-scale temperature reconstruction efforts since 1750 (Figure 6.1), it also highlighted that the divergence issue needed to be addressed at local/regional scales, as there can be individual variations and contributing factors particular to specific sites and their local conditions.

The divergence problem has potentially significant implications for the study of large-scale patterns of forest growth, the development of paleoclimatic reconstructions based on tree rings from northern forests, and understanding the global carbon cycle. The causes of this phenomenon, however, are not well understood and are difficult to test due to the existence of co-varying environmental factors that may potentially impact recent tree growth. Possible causes include the previously mentioned temperature-induced drought stress (Porter and Pisaric, 2011), nonlinear thresholds or time-dependent responses to recent warming, delayed snowmelt and related changes in seasonality, decreased snowpack and increased damaging spring frosts, and differential growth/climate relationships inferred for maximum, minimum, and mean temperatures (e.g., Wilson and Luckman, 2003; D'Arrigo et al., 2004; Wilmking et al., 2005; Juday, 2011; Wiles et al., 2012). Other possible explanations include local scale pollution (e.g., Wilson and Elling, 2004; Lloyd and Bunn, 2007; Rydval and Wilson, 2012), the solar dimming phenomenon, that can decrease the amount of solar radiation available for photosynthesis (A. Stine, Harvard University, pers. comm. 2012; D'Arrigo et al., 2008), and declines in stratospheric ozone levels that can increase incidence of ultraviolet radiation at the ground level (Briffa et al., 1998b; Briffa et al., 2004).

An additional consideration is that "end effects" and other methodological issues can emerge in standardization and chronology development, creating artifacts that may be misconstrued as divergence-type influences (e.g., Melvin and Briffa, 2010). Esper and Frank (2009) outline some of the pitfalls that can result in inadvertent detection of a divergence-type effect. These can occur during the detrending process, due to incomplete retention of low-frequency trends, sensitivity to the calibration period with temperature, or failure to account for the full season of tree growth response to climate. Further, divergence can only be reasonably identified in trees that have had a strong relationship to temperature at some point in the recent past. Although limited evidence suggests that the divergence may be anthropogenic in nature and restricted to the recent decades of the twentieth century (see Cook et al., 2004), more research, particularly regional studies, is needed to confirm these observations. In particular, a better understanding of the physiological ecology of tree growth at treeline could contribute significantly to resolving whether or not this phenomenon is taking place on a large scale, and aid identification of possible causes.

7

7. CONCLUSIONS AND FUTURE CHALLENGES

The northern tree-ring archive and related studies provide a long-term context with which to gauge the impacts of recent anthropogenic change relative to past natural variations. In this monograph, we have briefly reviewed four decades worth of TRL-LDEO northern tree-ring studies by the authors and colleagues (Figure 7.1) and the use of these tree-ring data to generate large-scale temperature reconstructions for the past millennium. Future challenges and outstanding issues remain, with important considerations and areas for improvement in the development of large-scale climatic reconstructions.

One key challenge is that temporal and spatial coverage of tree-ring data needs to be further improved in order to decrease uncertainty and enhance replication of chronologies and reconstructions, and to fill geographic gaps, not only in the tropics and Southern Hemisphere, but at the higher northern latitudes as well (e.g., in Scotland, central northern North America, Labrador).

Another consideration is the need to continue to refine our ability to capture low-frequency climate variability on multidecadal to centennial scales through tree-ring standardization, and via in-depth analyses of the use of RCS and similar detrending methods. These efforts require sampling of multiple age classes and strong replication (of trees per site and number of sites), particularly for millennial sites, if we are to decrease uncertainty during the MCA and other major climatic episodes.

As coverage improves, we will be better able to quantify the amplitude of trends back in time, notably the MCA and MCA-LIA transition as well as the timing of these periods. Improved coverage is also essential if we are to more precisely discern relative levels of natural and anthropogenic climatic changes, and to constrain climate sensitivity to anticipate anthropogenic impacts and future trends. Additional site sampling and continual updating of existing sites, as well as finding increased opportunities for the discovery of subfossil wood in northern settings (e.g., Wilson et al., 2011), so that ideally we would have millennial length chronologies every ~300 km or so across the northern latitudes, would substantially decrease the uncertainties in existing chronologies and reconstructions and provide better constrained spatial climate information.

Dendroclimatic Studies: Tree Growth and Climate Change in Northern Forests,
Monograph Number 67. Rosanne D'Arrigo, Nicole Davi, Gordon Jacoby,
Rob Wilson, and Greg Wiles.

Figure 7.1 Photographs of northern tree-ring field investigations and the authors. **A.** The very first tree (Siberian pine) cored in Mongolia (1475–1999) in 1995. The data from this tree and others at this site resulted in a publication on large-scale warming in Mongolia (Jacoby et al., 1996, *Science*). (Photo Credit: R. D'Arrigo) **B.** A cross section from a wood sample found in the Tarvagatay Mountains of Mongolia. The inner pith dates to 262 CE. (Photo Credit: N. Pederson) **C.** A Piper Cub plane getting ready to land in a make-shift runway of willow scrub in Wrangell St. Elias National Park and Preserve, Alaska. Other than trekking, a Piper Cub is the only way into the park. (Photo Credit: N. Davi)

Figure 7.1 ***(Cont'd)*** **D.** A tree core from Alaska showing a VERY NARROW, light latewood band at the year 1783. The Laki volcanic eruption occurred in Iceland during that year (in June) and cooled global temperatures, causing this tree to stop growing abruptly in the fall of that year (Jacoby et al., 1999). (Photo Credit: G. Jacoby) **E.** Dr. Jacoby stands on top of ancient White River Ash, a layer of ash from Mt. Churchill that blanketed more than 340,000 sq km of eastern Alaska and NW Canada in 800AD. Tree stumps buried by the event were exposed as the bank eroded. These samples, along with living white spruce, were used to extend our understanding of climate back by nearly two thousand years. (Photo Credit: N. Davi) **F.** 400-year-old Siberian larch trees growing in the Mongolian steppe. These trees are growing in very arid conditions and are sensitive to changes in moisture. (Photo Credit: B. Buckley)

G

H

Figure 7.1(*Cont'd*) G. Davi in a white spruce forest overlooking Mendenhall Glacier in Juneau, AK. These trees were used to learn about climate change and glacier dynamics over the past 400–500 years. (Photo Credit: G. Jacoby) **H.** Drs. Gordon Jacoby and Rosanne D'Arrigo sampling native trees in Alaska in the early 1980s. (Photo Credit: Unknown)

I

J

Figure 7.1(*Cont'd*) I. Dr. Greg Wiles sampling trees near tree line in Glacier Bay National Park and Preserve, Alaska. (Photo Credit: Unknown) **J.** Rob Wilson beside relict root stocks (preserved in peat) of ancient Scots pine trees in Glen Affric (Scottish highlands) at about 550 m above current treeline. As yet, this location is not radiocarbon dated, but a similar site in the Cairngorms (90 km to east) has been dated to ~6000 years BP. (Photo Credit: A. Wilson). For color detail, please see color plate section.

Selecting sites using criteria that tend to maximize the occurrence of temperature-sensitive trees and screening for chronology quality are also essential in order to optimize the climate signal in the northern tree-ring data, something that requires informed discernment by skilled dendrochronologists (Frank et al., 2010). Such an approach is emphatically not "cherry picking," but rather an informed strategy aimed at maximizing the skill of resulting reconstructions. Such screening is essential in order to select only those sites and chronologies that reflect the climate variable of interest.

Beyond the large-scale reconstructions, focus on regional climatic changes and projections, a key focus of the IPCC, may be of equal or greater relevance to policymakers (e.g., Lenton, 2011). It is important to be aware that uncertainties in instrumental data also exist and must be minimized or corrected wherever possible (e.g., Brohan et al., 2006; Frank et al., 2007; Thompson et al., 2008). This is particularly the case in remote northern areas where sparse and short station records typically limit calibration and verification testing of reconstruction models. This inevitable fact results in an underestimation, in many cases, of the strength of the climate signal in the tree-ring data. On the positive side, recent efforts on the reconstruction and compilation of historical observations (e.g., atmospheric circulation reconstructions over the earth (ACRE), led by Rob Allan, Project Manager, http://www.met-acre.org/Home/manager) have much potential to provide longer records for reconstruction model testing and development.

Refinement of new methodologies (e.g., tree growth modeling and proxy data-model comparisons, pseudoproxy analyses [Glossary], climatic field reconstructions [CFR], Bayesian techniques, ensemble reconstructions, and challenge-type experiments) (e.g., Frank et al., 2010; Graham et al., 2011) allows us to quantitatively compare proxies with climate models to help decrease inherent noise in these data sets and help constrain climate sensitivity to doubled CO_2 concentrations (e.g., Hegerl et al., 2006), and is a key area for future research. Ensemble/probabilistic assessments of reconstructions can be highly useful in estimating uncertainties in temperature amplitudes. The radiative forcing time series used in climate models needs further refinement in this regard to improve model hindcasts and predictions, as does the understanding of internal atmosphere-ocean climate dynamics. CFRs are a key element in the shift to the "ideal" of spatial reconstructions and the successful application of tree rings and other proxies to increase understanding of atmosphere-ocean climate variability, and allow comparison with climate models that provide dynamical and physical context. Forward modeling methods should considerably improve efforts to understand causes of divergence-type effects.

Aforementioned efforts to compare recent and MCA warming suggest how extreme recent warming has been relative to the natural fluctuations of the past millennium. This conclusion, however, must be taken cautiously. Firstly, there is significant divergence between reconstructed and actual temperatures since the mid-1980s. Secondly, there are presently only very few millennial-length records available for direct comparison between the recent period and the MCA, and these records show trends that are not necessarily coherent over the latter interval. Ultimately, many long records from new locations across the northern latitudes and updating of existing records to the present are required. It is particularly important to increase the development of millennial and submillennial length MXD records, which can be processed from already-collected wood samples as well as new investigations.

Successful modeling of paleoclimate data with the high temperatures of the late 1990s is essential if we are to make robust, definitive conclusions about past temperature amplitudes and variability. Refined data-model comparisons (e.g., NOAA Paleoclimatology Challenge, Wahl et al., 2010) will allow us to better predict future conditions in the earth's climate system. Process modeling methods that combine tree rings or other proxies with simulated physical conditions and climate model output (e.g., so-called Proxy Surrogate Reconstructions or PSR) are providing new understanding of climate variability (e.g., Graham et al., 2010; Trouet et al., 2009).

Finally, assumptions of linearity, stationarity, and uniformitarianism are overly simplistic in a world of rapid anthropogenic change. We must thus be increasingly aware of such issues as divergence, possible CO_2 or nutrient fertilization, changing water use efficiency, extreme climatic events, pollution effects, logging and development, and increasing disturbance due to rapid warming (insects, fire, and disease), and their possible effects on forest growth. Rapid warming and other factors are also modifying tree growth seasonality and the climatic response of tree-ring proxies. Multiproxy data comparisons (e.g., Moberg et al., 2005; Ljungqvist, 2010) can yield a greater range of resolution to climate but require caution due to the inherent differences and incompatibilities among proxies. Independent use of proxies for validation purposes is likely more effective in testing for consensus among such archives. Finally, improvements in communication between the paleoclimatic, observational, and modeling communities are still very much needed.

The findings of the nearly four decades of research by the investigators at the TRL-LDEO have been integrated herein into a single synthesis product of their northern forest temperature-related tree-ring studies. The resulting compilation has yielded a valuable long-term perspective on this research, and will provide insights into future studies by the investigators and others, as the process of dendroclimatic applications in the earth sciences continues to evolve. We anticipate that this research and dataset will be of considerable value for educational purposes as part of the ongoing process of research and scholarship within the scientific community.

GLOSSARY

Age-Band Decomposition (ABD): A method of standardization that helps preserve low-frequency variability in tree rings (Briffa et al. 2001). The rationale is that this method avoids the need to detrend individual tree core time series to remove effects of tree age, allowing stronger variance on longer time scales. This approach, an extension of the Regional Curve Standardization method, includes dividing the dataset into different age classes prior to processing.

Arctic Oscillation (AO): The dominant pattern of non-seasonal sea-level pressure variations north of 20°N latitude, characterized by pressure anomalies of one sign in the Arctic with the opposite anomalies centered about 37–45°N (Thompson and Wallace, 1998). The AO is a close relative of the North Atlantic Oscillation (NAO), although there is some discussion regarding which of these is more physically meaningful with regard to atmospheric dynamics. The AO impacts climate over large areas of the NH, include Europe and North America.

Autocorrelation: The tendency for memory in observations as a function of the time between them. For example, ring-widths in treeline conifers are often autocorrelated due to cambial processes, which can cause a more ambiguous temperature signal at higher frequencies, as can occur following volcanism (see text; Fritts, 1976; Cook and Kairiukstis, 1990; D'Arrigo et al., 2009).

Climate Field Reconstructions (CFR): A multivariate technique of paleoclimatic reconstruction in which spatial patterns of climate are reconstructed using proxy and instrumental data, providing an added dimension to point reconstructions of single time series (e.g., Zhang et al., 2004).

CO_2 Fertilization: The theory, currently being debated, that trees will respond directly to increasing concentrations of carbon dioxide in the atmosphere by enhancing their radial growth (LaMarche et al., 1984).

Cross Dating: The basic principle of dendrochronology, i.e., the matching of patterns of narrow and wide rings between tree core samples of trees of the same

Dendroclimatic Studies: Tree Growth and Climate Change in Northern Forests,
Monograph Number 67. Rosanne D'Arrigo, Nicole Davi, Gordon Jacoby,
Rob Wilson, and Greg Wiles.

species at the same site, which ensures precise annual dating of a tree-ring series or chronology (Fritts, 1976). The cross dating of site chronologies can be verified against neighboring site chronologies of the same species that have been independently dated.

Dendroclimatology: The application of the science of tree-ring analysis, or dendrochronology, to the understanding and reconstruction of past climate variability over the past several centuries to millennia.

Divergence Problem: The observed decoupling of tree growth and temperature observed at some northern latitude tree-ring sites over recent decades, in such a way that tree growth estimates of climate appear to underestimate temperature (D'Arrigo et al., 2008).

End Effects: Biases that can occur at the beginning and end of tree-ring time-series that are related to an artifact of the tree-ring standardization process (Briffa and Cook, 2008; Melvin and Briffa, 2008; Cook and Kairiukstis, 1990).

ENSO: The El Niño-Southern Oscillation: A quasi-periodic pattern of tropical Pacific atmosphere-ocean climate, which influences weather and climate throughout the world through teleconnections (Trenberth et al., 2007).

False Rings: False rings are bands of what appear to be latewood cells, followed by earlywood and then by true latewood in annual growth rings (Fritts, 1976; Cook and Kairiukstis, 1990). Such false rings tend to occur due to environmental stress (e.g., drought) during the growing season, and are potentially problematic in cross dating of tree-ring series at sites where they occur, although, for most species, they are typically readily identifiable via the cross-dating process.

Forward Modeling: Mechanistic models of tree-ring formation used to simulate tree growth response to critical climate variables such as temperature and/or precipitation (e.g., Anchukaitis et al., 2006).

Frost Rings: Anatomical cellular damage in annual growth rings in wood that can occur during adverse climatic conditions, such as cold associated with major volcanic events (LaMarche and Hirschboeck, 1984).

ITRDB: International Tree-Ring Data Bank, part of the NOAA National Climatic Data Center http://www.ncdc.noaa.gov/paleo/treering.html.

Light Rings: Tree rings with very thin and pale latewood. Such low-density rings are often associated with volcanic or other adverse events (Filion et al., 1986).

Little Ice Age: A time period from approximately 1450–1850 CE, during which colder conditions were observed in many areas of the globe, particularly the North Atlantic region (Grove, 1988).

Maximum Latewood Density (MXD): A parameter that measures peak density of the latewood cells in a given tree ring (Schweingruber, 1988). MXD is sensitive to warm season temperatures and is complementary to ring width.

Medieval Climate Anomaly: A time period early in the past millennium (ca. 950–1250 AD) during which many regions of the globe had warmer temperatures based on proxy data (Diaz et al., 2011).

Missing Rings: Annual growth that is missing for some, but not all, trees and wood samples from a given site. Such missing rings (also sometime referred to as locally absent rings) are accounted for in the process of cross dating (Stokes and Smiley, 1968). The phenomenon of a stand-wide missing ring in a given year has never been correctly documented to occur in actual tree-ring data (RW/MXD) among *all* trees at a given northern treeline location.

North Atlantic Oscillation (NAO): An alternation in atmospheric mass (sea-level pressure difference between the Arctic and the subtropical Atlantic) that is the most prominent and recurrent pattern of atmospheric variability in the middle to higher latitudes of the NH, and which is associated in its positive and negative phases with climate anomalies in the circum-North Atlantic region (Hurrell et al., 2003).

Northern Treeline: The high latitudinal extent of trees in northern regions where low summer temperatures limit growth, which makes these trees very sensitive recorders of past temperature variability.

Polar Amplification: The theory that the higher northern latitudes are warming faster than other regions of the globe due to positive feedback factors (e.g., increased ice melt), which decreases the albedo, causing further warming.

Pseudoproxy Reconstructions: Synthetic data sets can be generated by adding artificial noise to either actual instrumental climate records or model simulations. These pseudo-data sets are used to develop pseudoproxy reconstructions, which can then be compared to the original record or simulation (Smerdon, 2012). This approach is an invaluable approach to testing the robustness of different reconstruction methodologies.

Regional Curve Standardization (RCS): A method of standardization of raw tree-ring measurements (Briffa and Melvin, 2010) that can enhance the potential retention of low-frequency climatic information in tree-ring data at time scales greater than the mean length of the tree-ring series (Cook et al., 1995).

Signal-Free Standardization: A method of tree-ring standardization that aims to prevent trend distortion, a type of end-effect bias, resulting from the external influence of the actual climate signal, which biases the biological growth trend (Melvin and Briffa, 2008).

Standardization: The processing of raw ring-width or density measurements from wood samples that removes the effect of biological age trend on the resultant tree-ring chronology (Cook, 1985). Standardization is also often referred to as detrending.

Subfossil Wood: Preserved dead wood material that can be used, through overlap with the wood from living trees and historical structures, to extend living tree-ring records back in time (Fritts, 1976).

Tree-Ring Reconstructions: Tree-ring-based time series that extends the available instrumental record of climate into the past through the generation of a transfer function that calibrates the tree-ring data with the instrumental record.

X-Ray Densitometry: The process of generating density time series from conifer tree rings (e.g., Schweingruber, 1988; Thetford et al., 1991). This typically involves a scanning method coupled to an image analysis system. Recently, the blue intensity (BI) parameter has been shown to portray similar properties to MXD (Campbell et al., 2007; Wilson et al., 2011).

Vascular Cambium: A layer of meristematic cells found in plants that undergo secondary growth, which divides to produce secondary xylem to the inside of the plant and secondary phloem to the outside.

References Cited

Anchukaitis, K., M. Evans, A. Kaplan, E. Vaganov, M. Hughes, H. Grissino-Mayer, and M. Cane (2006), Forward modeling of regional-scale tree-ring patterns in the southeastern United States and the recent influence of summer drought. *Geophys. Res. Lett.*, 33: L04705, doi:10.1029/2005GL025050.

Briffa K., T. Osborn, F. Schweingruber, I. Harris, P. Jones, S. Shiyatov, and E. Vaganov (2001), Low-frequency temperature variations from a northern tree-ring-density network. *Journal of Geophysical Research* 106: 2929–2941.

Briffa, K., and E. Cook (2008), What are the sources of uncertainty in the tree-ring data: how can they be quantified and represented? White paper presented as part of paleoclimatic meeting on proxy uncertainties, held in Trieste, Italy, June 9–11.

Briffa, K., and T. Melvin (2010), A closer look at Regional Curve Standardisation of tree-ring records: justification of the need, a warning of some pitfalls, and suggested improvements in its application. In: Hughes, M.K., Diaz, H.F., Swetnam, T.W., Eds. Dendroclimatology: Progress and Prospects. Springer Verlag.

Campbell R., D. McCarroll, N. Loader, H. Grudd, I. Robertson, and R. Jalkanen (2007), Blue intensity in *Pinus sylvestris* tree rings: Developing a new palaeoclimate proxy. *The Holocene* 17(6): 821–828.

Cook, E. (1985), A time series analysis approach to tree-ring standardization. Ph.D. thesis, University of Arizona.

Cook, E., and L. Kairiukstis (1990), Methods of Dendrochronology. Kluwer, Dordrecht.

Cook, E., K. Briffa, D. Meko, D. Graybill, and G. Funkhouser (1995), The 'segment length curse' in long tree-ring chronology development for paleoclimatic studies. *The Holocene* 5: 229–237.

D'Arrigo, R., R. Wilson, B. Liepert, and P. Cherubini (2008), On the 'Divergence Problem' in northern forests: A review of the tree-ring evidence and possible causes. *Global and Planetary Change* 60: 289–305. doi:10.1016/j.gloplacha.2007.03.004.

D'Arrigo, R., R. Wilson, and A. Tudhope (2009), Impact of volcanic forcing on tropical climate during the past five centuries. *January. Nature Geoscience* doi:10.1038/NGE0393.

Diaz, H., R. Trigo, M. Hughes, M. Mann, E. Xoplaki, and D. Barriopedro (2011), Spatial and temporal characteristics of climate in Medieval times revisited. *Bull. Amer. Met. Soc*. November, 1487–1500.

Filion, L., S. Payette, L. Gauthier, and Y. Boutin (1986), Light rings in subarctic conifers as a dendrochronological tool. *Quaternary Research*, 26: 272–279.

Fritts, H. (1976), Tree Rings and Climate. Academic Press, NY.

Grove, J. 1988. The Little Ice Age. Routledge, NY.

Hurrell, J., Y. Kushnir, G. Ottersen, and M. Visbeck. (2003), The North Atlantic Oscillation: Climatic significance and environmental impact. *AGU Monograph* Series 134: 279 pp. ISSN: 0065-8448; ISBN: 0-87590-994-9.

LaMarche, V., and K. Hirschboeck (1984), Frost rings in trees as records of major volcanic eruptions. *Nature* 307: 121–126.

LaMarche, V., D. Graybill, H. Fritts, and M. Rose (1984), Increasing atmospheric carbon dioxide: tree-ring evidence for growth enhancement in natural vegetation. *Science* 225: 1019–1021.

Melvin, T., and K. Briffa (2008), A "signal-free" approach to dendroclimatic standardisation. *Dendrochronologia* 26: 71–86.

Schweingruber, F. (1988), Tree Rings. Basics and Applications of Dendrochronology. Kluwer, Dordrecht.

Smerdon, J. (2012), Climate models as a test bed for climate reconstruction methods: pseudoproxy experiments. *WIREs Climate Change* 3: 63–77.

Stokes, M., and T. Smiley (1968), An Introduction to Tree-Ring Dating. University of Arizona Press.

Thompson, D., and J. Wallace (1998), The Arctic oscillation signature in the wintertime geopotential height and temperature fields. *Geophysical Research Letters* 25 (9): 1297–1300. Bibcode 1998GeoRL..25.1297T. doi:10.1029/98GL00950.

Trenberth, K., P. D. Jones, P. Ambenje, R. Bojariu , D. Easterling, A. Klein Tank, D. Parker, F. Rahimzadeh, J. A. Renwick, M. Rusticucci, B. Soden, and P. Zhai (2007), "Observations: Surface and Atmospheric Climate Change." In: Solomon, S., D. Qin, M. Manning, Z. Chen, M. Marquis, K. B. Averyt, M. Tignor, and H. L. Miller. Climate Change 2007: The Physical Science Basis. Contribution of Working Group I to the Fourth Assessment Report of the Intergovernmental Panel on Climate Change. Cambridge, UK: Cambridge University Press. pp. 235–336.

Zhang, Z., M. Mann, and E. Cook (2004), Alternative methods of proxy-based climate field reconstruction: application to summer drought over the conterminous United States back to AD 1700 from tree-ring data. *The Holocene* 14: 502–516.

REFERENCES

Ababneh, L. (2006), Analysis of radial growth patterns of strip-bark and whole-bark bristlecone pine trees in the White Mountains of California: implications in paleoclimatology and archaeology of the Great Basin, Ph.D. thesis, University of Arizona, Tucson.

ACIA (Arctic Climate Impact Assessment) (2005), Impacts of a Warming Arctic. Cambridge University Press, New York.

Anchukaitis, K., M. Evans, A. Kaplan, E. Vaganov, M. Hughes, H. Grissino-Mayer, and M. Cane (2006), Forward modeling of regional-scale tree-ring patterns in the southeastern United States and the recent influence of summer drought, *Geophys. Res. Lett.*, 33, L04705, doi:10.1029/2005GL025050, 2006.

Baillie, M. G. L. (1999), Exodus to Arthur: catastrophic encounters with comets, Batsford, London.

Banfield, C. E., and J. Jacobs, (1998), Regional patterns of temperature and precipitation for Newfoundland and Labrador during the past century, *The Canadian Geographer*, 42 (4).

Barber, V., G. Juday, and B. Finney (2000), Reduced growth of Alaska white spruce in the twentieth century from temperature-induced drought stress, *Nature*, 405, 668–672.

Barber, V., G. Juday, T. Ostercamp, R. D'Arrigo, E. Berg, B. Buckley, L. Hinzman, H. Huntington, T. Jorgensen, D. McGuire, B. Riordan, A. Whiting, G. Wiles, and M. Wilmking. 2009. Chapter 9: A synthesis of recent climate warming effects on terrestrial ecosystems of Alaska, In: *Climate Warming in Western North America/Evidence and Environmental Effects*, F. H. Wagner, Ed., University of Utah Press.

Barclay, D. J., G. Wiles, and P. Calkin (1999), A 1119-year tree-ring width chronology from western Prince William Sound, southern Alaska, *The Holocene*, 9, 79–84.

Beck, P., G. Juday, C. Alix, V. Barber, S. Winslow, E. Sousa, P. Heiser, J. Herriges, and S. Goetz (2011), Changes in forest productivity across Alaska consistent with biome shift, *Ecological Research Letters*, doi:10.1111/j.1461-0248.2011.01598.x.

Beck, P., L. Andreu-Hayles, R. D'Arrigo, K. J. Anchukaitis, C. Tucker, J. Pinzon, and S. Goetz (2013), A large-scale coherent signal of temperature and canopy status in maximum latewood density of tree rings at Arctic treeline in North America, *Envir. Res. Lett.*, 100, 109–118, doi:10.1016/j.gloplacha.2012.10.005.

Biondi, F., A. Gershunov, and D. Cayan (2001), North Pacific decadal climate variability since 1661, *J. Climate*, 14, 5–10.

Dendroclimatic Studies: Tree Growth and Climate Change in Northern Forests,
Monograph Number 67. Rosanne D'Arrigo, Nicole Davi, Gordon Jacoby,
Rob Wilson, and Greg Wiles.

Bonan, G. (2008), Forests and climate change: forcings, feedbacks and the climate benefits of forests, *Science*, 320, 1444–1449.

Brienen, R., E. Gloor, and P. Zuidema (2012), Detecting evidence for CO_2 fertilization from tree ring studies: the potential role of sampling biases of selection of forest trees. *Global Biogeoch. Cycles*, 26, GB1025, 13 pp., 2012 doi:10.1029/2011GB004143.

Briffa, K. (2000), Annual climate variability in the Holocene: Interpreting the message from ancient trees, *Quat. Sci. Rev.*, 19, 87–105.

Briffa, K., and T. Melvin (2010), A closer look at Regional Curve Standardisation of tree-ring records: justification of the need, a warning of some pitfalls, and suggested improvements in its application, In: Hughes, M. K., Diaz, H. F., Swetnam, T. W. (Eds.) *Dendroclimatology: Progress and Prospects*. Springer Verlag.

Briffa, K., P. Jones, T. Bartholin, D. Eckstein, F. Schweingruber, W. Karlen, P. Zetterberg, and M. Eronen (1992), Fennoscandian summers from AD 500: temperature changes on short and long time scales, *Clim. Dyn.*, 7, 111–119.

Briffa, K., P. Jones, F. Schweingruber, and T. Osborn (1998a), Influence of volcanic eruptions on Northern Hemisphere summer temperature over the past 600 years, *Nature*, 393, 450–455.

Briffa, K., F. Schweingruber, P. Jones, and T. Osborn (1998b), Reduced sensitivity of recent tree growth to temperature at high northern latitudes, *Nature*, 391, 678–682.

Briffa, K., F. Schweingruber, P. Jones, T. Osborn, I. Harris, S. Shiyatov, A. Vaganov, and H. Grudd (1998c), Trees tell of past climates: but are they speaking less clearly today? *Phil. Trans. R. Soc. Lond. B.*, 353, 65–73.

Briffa, K.R., T. Osborn, F. Schweingruber, I. Harris, P. Jones, S. Shiyatov, and E. Vaganov (2001), Low frequency temperature variations from a northern tree-ring density network, *J. Geophys, Res.*, 106, 2929–2941.

Briffa, K., T. Osborn, and F. Schweingruber (2004), Large-scale temperature inferences-from tree rings: a review, *Global and Planetary Change*, 40, 11–26.

Brohan, P., J. Kennedy, I. Harris, S. Tett, and P. Jones (2006), Uncertainty estimates in regional and global observed temperature changes: A new data set from 1850, *J. Geophys. Res.*, 111, D12106, doi:10.1029/2005JD006548.

Bunn, A., and S. Goetz (2006), Trends in satellite-observed circumpolar photosynthetic activity from 1982 to 2003: the influence of seasonality, cover type, and vegetation density, *Earth Interact.*, 10, Paper No. 12.

Büntgen, U., D. Frank, D. Nievergelt, and J. Esper (2006), Summer temperature variations in the European Alps, AD 755–2004. *J. Climate*, 19, 5606–5623.

Büntgen, U., W. Tegel, K. Nicolussi, M. McCormick, D. Frank, V., Trouet, J. Kaplan, F. Herzig, K. Heussner, H. Wanner, J. Luterbacher, and J. Esper (2011), 2500 years of European climate variability and human susceptibility, *Science*, 331, 578–582.

Chapin, F.S. III, T. Callaghan, Y. Bergeron, M. Fukuda, J. Johnstone, G. Juday, and S. Zimov (2004), Global change and the boreal forest: shifting states or gradual change? *Ambio*, 33, 361–365.

Christiansen, B., T. Schmith, and P. Thejll (2009), A surrogate ensemble study of climate reconstruction methods: stochasticity and robustness, *J. Climate*, 22, 951–976.

Christiansen, B., and F. C. Ljungqvist (2012), The extra-tropical Northern Hemisphere temperature in the last two millennia: reconstructions of low-frequency variability, *Climate of the Past*, 8, 765–786.

Cook, E. (1985), A time series analysis approach to tree-ring standardization, Ph.D. Thesis, University of Arizona.

Cook, E., and L. Kairiukstis (1990), Methods of Dendrochronology, Kluwer, Dordrecht.

Cook, E., and P. Krusic (2004), The North American Drought Atlas, http://iridl.ldeo.columbia.edu/SOURCES/.LDEO/.TRL/.NADA2004/.pdsi-a. http://www.ncdc.noaa.gov/paleo/pdsi.html.

Cook, E., and K. Peters (1997), Calculating unbiased tree-ring indices for the study of climate and environmental change, *The Holocene*, 7, 361–370.

Cook, E. R., K. J. Anchukaitis, B. M. Buckley, R. D. D'Arrigo, G. C. Jacoby, and W. E. Wright (2010), Asian monsoon failure and megadrought during the last millennium, *Science*, 328(5977), 486–489, doi:10.1126/science.1185188.

Cook, E., K. Briffa, D. Meko, D. Graybill, and G. Funkhouser (1995), The "segment length curse" in long tree-ring chronology development for palaeoclimatic studies, *The Holocene*, 5, 229–237.

D'Arrigo, R., and R. Wilson (2006), On the Asian expression of the PDO, *Internat. J. Climatol.*, August 2006: doi:10.1002/joc.1326.

Diaz, H., R. Trigo, M. Mann, R. Garcia-Herrera, M. Hughes, and E. Xoplaki (2010), Proceedings, Workshop on the Medieval Climate Anomaly (MCA), Convenors. 22–24 September, 2010, Lisbon, Portugal.

Esper, J., and D. Frank (2009), Divergence pitfalls in tree-ring research, *Climatic Change*, 94, 261–266.

Esper, J., and F. Schweingruber (2004), Large-scale treeline changes recorded in Siberia, *Geophys. Res. Lett.*, 31, doi:10.1029/2003GL019178.

Esper, J., E. Cook, and F. Schweingruber (2002), Low-frequency signals in long tree-ring chronologies for reconstructing past temperature variability, *Science*, 295, 2250–2253.

Esper, J., E. Cook, P. Krusic, and K. Peters (2003), Tests of the RCS method for preserving low-frequency variability in long tree-ring chronologies, *Tree-Ring Research*, 59, 81–98.

Esper, J., D. Frank, D., and R. Wilson (2004), Climate reconstructions: Low-frequency ambition and high-frequency ratification, *EOS*, 85(12).

Esper, J., R. Wilson, D. Frank, A. Moberg, H. Wanner, and J. Luterbacher (2005), Climate: past ranges and future changes, *Quat. Sci. Rev.*, 24, 2164–2166.

Esper J., D. Frank, U. Büntgen, A., Verstege, R. Hantemirov, and A. Kirdyanov (2010), Trends and uncertainties in Siberian indicators of 20th century warming, *Global Change Biology*, 16, 386–398.

Esper, J., D. Frank, M. Timonen, E. Zorita, R. Wilson, J. Luterbacher, S. Holzkamper, N. Fischer, S. Wagner, D. Nievergelt, A. Verstege, and U. Büntgen (2012), Orbital forcing of tree-ring data, *Nature Climate Change*, doi:10.1038/nclimate1589.

Esper, J., L. Schneider, P. Krusic, J. Luterbacher, U. Büntgen, M. Timonen, F. Sirocko, and E. Zorita (2013), European summer temperature response to annually dated volcanic eruptions over the past nine centuries, *Bulletin of Volcanology*, 75, 736, doi:10.1007/s00445-013-0736-z.

Esper J., U. Büntgen, J. Luterbacher, and P. Krusic (in press), Testing the hypothesis of post-volcanic missing rings in temperature sensitive dendrochronological data, *Dendrochronologia*.

Evans, M., B. Reichert, A. Kaplan, K. Anchukaitis, E. Vaganov, M. Hughes, and M. Cane (2006), A forward modeling approach to paleoclimatic interpretation of tree-ring data, *Geophys. Res. Biogeosciences*, 111, G03008, doi:10.1029/2006JG000166.

Frank, D., and J. Esper (2005), Temperature reconstructions and comparisons with instrumental data from a tree-ring network for the European Alps, *Int. J. Climatol.*, 25, 1437–1454.

Frank, D., J. Esper, E. Zorita, and R. Wilson (2010), A noodle, hockey stick and spaghetti plate—a perspective on high resolution paleoclimatology, *Wiley Interdisciplinary Reviews on Climate Change*, 1, 507–516.

Frank, D., U. Büntgen, R. Bohm, M. Maugeri, and J. Esper (2007), Warmer early Instrumental measurements versus colder reconstructed temperatures: shooting at a moving target, *Quat. Sci. Rev.*, 26, 3298–3310.

Frank, D., J. Esper, and E. Cook (2007), Adjustment for proxy number and coherence in a large-scale temperature reconstruction, *Geophys. Res. Lett.*, 34, L*16709*, doi:10.1029/2007GL030571.

Fritts, H. (1976), Tree Rings and Climate, Academic Press, NY.

Giddings, J. (1941), Dendrochronology in northern Alaska, *Univ. Ariz. Bull.*, 12 (4).

Goldstein, G.H. (1981), Ecophysiological and Demographic Studies of White Spruce (*Picea glauca* [Moench] Voss) at Treeline in the Central Brooks Range of Alaska, Ph.D. Thesis, University ofWashington, Seattle.

Goose, H. (2010), Degrees of climate feedback, *Nature*, 463, 438–439.

Graham, N., C. Ammann, D. Fleitmann, K. Cobb, and J. Luterbacher (2010), Support for global climate reorganization during the 'Medieval Climate Anomaly," *Climate Dynamics*, doi:10.1007/s00382-010-0914-z.

Graham, N., R. D'Arrigo, K. Anchukaitis, E. Wahl, and D. Anderson (2011), Paleoclimate Reconstruction Challenge, IGBP PAGES/World Data Center for Paleoclimatology, NOAA/NCDC Paleoclimatology Program, Boulder CO, USA.

Graumlich, L., and J. King (1997), Late Holocene climatic variation in northwestern Alaska as reconstructed from tree rings, A Final Report on Cooperative Research with the National Park Service.

Graybill, D., and S. Idso (1993), Detecting the aerial fertilization effect of atmospheric CO_2 enrichment in tree-ring chronologies, *Global Biogeochem. Cycles*, 7, 81–95.

Grebmeier, J. M., J. Priscu, R. D'Arrigo, H. Ducklow, C. Fleener, K. Frey, and C. Rosa (2011), National Research Council of the National Academies Report: Frontiers in Understanding Climate Change and Polar Ecosystems, The National Academies Press: Washington D.C., 84 pp.

Grove, J. (1988), The Little Ice Age, Academic Press, London.

Groveman, B. S., and H. E. Landsberg (1979), Simulated northern hemisphere temperature departures 1579–1880, *Geophys. Res. Lett.*, 6, 767–770.

Grudd, H. (2008), Torneträsk tree-ring width and density AD 500–2004: A test of climatic sensitivity and a new 1500-year reconstruction of north Fennoscandian summers, *Climate Dynamics*, 31, 843–857. doi:10.1007/s00382-007-0358-2.

Hansen, J., R. Ruedy, M. Sato, and K. Lo. (2010), Global surface temperature change, *Rev. Geophys.*, 48, RG4004, doi:10.1029/2010RG000345.

Hantemirov, R.M., and S. Shiyatov (2002), A continuous multimillennial ring-width chronology in Yamal, northwestern Siberia,*The Holocene*, 12 (6), 717–726.

Havranek, W. M., and W. Tranquillini (1995), Physiological processes during winter dormancy and their ecological significance, pp. 95–124 In: Ecophysiology of Coniferous Forests, Edited by W.K. Smith and T.M. Hinckley, Academic Press, New York.

Hegerl, G., T. Crowley, W. Hyde, and D. Frame (2006), Climate sensitivity constrained by temperature reconstructions over the past seven centuries, *Nature*, 440, 20 April 2006; doi:10.1038/Nature 04679.

Hegerl, G. C., T. Crowley, M. Allen, W. T. Hyde, H. Pollack, J. Smerdon, and E. Zorita (2007), Detection of human influence on a new 1500 yr climate reconstruction, *J. Climate*, 20, 650–666.

Hickler, T., B. Smith, I. C. Prentice, K. Mjofors, P. Miller, A. Arneth, and M. Sykes (2008), CO_2 fertilization in temperate FACE experiments not representative of boreal and tropical forests, *Global Change Biology*, 14, 1531–1542.

Hughes, M. K., E. Vaganov, S. Shiyatov, R. Touchan, and G. Funkhouser (1999), Twentieth-century summer warmth in northern Yakutia in a 600 year context, *The Holocene*, 9, 603–608.

Jansen, E., J. Overpeck, K. R. Briffa, J.-C. Duplessy, F. Joos, V. Masson-Delmotte, D. Olago, B. Otto-Bliesner, W. R. Peltier, S. Rahmstorf, R. Ramesh, D. Raynaud, D. Rind, O. Solomina, R. Villalba, and D. Zhang (2007), Intergovernmental Panel on Climatic Change (IPCC). The Physical Science Basis. Summary for Policy Makers (Solomon, S., D. Qin, M. Manning, Z. Chen, M. Marquis, K. B. Avery, M. Tignor, and H. L. Miller, Eds.), Cambridge University Press, Cambridge, United Kingdom and New York, NY, USA.

Jones, P., and M. E. Mann (2004), Climate over past millennia, *Rev. Geophys.*, 42, RG2002, doi:10.1029/2003RG000143.

Jones, P., K. Briffa, T. Barnett, and S. Tett (1998), High-resolution paleoclimatic records for the last millennium: interpretation, integration and comparison with General Circulation Model control-run temperatures, *The Holocene*, 8, 455–471.

Juday, G. (2011), The biome shift now occurring in the boreal region: characteristics and a first look at knowledge needs, Presentation, Frontiers in Understanding Climate Change and Ecosystems, Workshop, August 2010, Cambridge, Maryland.

Juday, G. P., V. Barber, E. Vaganov, S. Rupp, S. Sparrow, J. Yarie, and H. Linderholm, Lead Authors.E. Berg, R. D'Arrigo, P. Duffy, O. Eggertsson, V. Furyaev, E. Hogg, S. Huttunen, G. Jacoby, V. Kaplunov, S. Kellomaki, A. Kirdyanov, C. Lewis, S. Linder, M. Naurzbaev, F. Pleshikov, U. Runesson, Y. Savva, O. Sidorova, V. Stakanov, N. Tchebakova, E. Valendik, E. Vedrova, and M. Wilmking,Consulting Authors. (2005), Forests, Land Management, Agriculture, Chapter 14, pp. 781–862, In: Arctic Climate Impact Assessment (ACIA). Arctic Council. Cambridge University Press.

Kozlowski, T. T. (1971), Growth and Development of Trees, II. Cambial Growth, Root Growth, and Reproductive Growth, Academic Press, New York.

Labrador Highlands Research Group (2005), Annual Report of Research 2004. Labrador Highlands Research Group, Department of Geography, Memorial University, St-John's NL A1B 3X9.M.

Larsen, L., B. Vinther, K. Briffa, T. Melvin, H. Clausen, P. Jones, S. Andersen, C. Hammer, M. Eronen, H. Grudd. R. Hantemirov, M. Naurzbaev, and K. Nicolussi (2008), New ice core evidence for a volcanic cause of the A.D. 536 dust veil, *Geophys. Res. Lett.*, 35, L04708, doi:10.1029/2007GL032450.

Lenton, T. (2011), 2 °C or not 2 °C? That is the climate question, *Nature*, 473, 7.

Lloyd, A. H., and A. G. Bunn (2007), Responses of the circumpolar boreal forest to 20th century climate variability, *Environmental Res. Lett.*, 2, 045013, doi:10.1088/1748–9326/2/4/045013.

Lloyd, A., and C. Fastie (2002), Spatial and temporal variability in the growth and climate response of treeline trees in Alaska, *Climatic Change*, 58, 481–509.

Lloyd, A., and C. Fastie (2003), Patterns and dynamics of treeline advance on the Seward Peninsula, Alaska, *J. Geophys. Res.-Atmospheres*, 108 (D2), 8161, doi:10.1029/2001JD000852.

Luckman, B. H. (2010), Tree Rings as Temperature Proxies, Geoscience of Climate and Energy 6, *Geoscience Canada*, 37,1, 38–42.

Luckman, B. H., and R. Wilson (2005), Summer temperature in the Canadian Rockies during the last millennium? a revised record, *Climate Dynamics*, 24, 131–144. doi:10.1007/s00382-004-0511-0.

Ljungqvist, F. C. (2010), A new reconstruction of temperature variability in the extra-tropical Northern Hemisphere during the last two millennia, Geografiska Annaler: Series A, *Physical Geography*, 92, 339–351.

Mann, M. E., R. Bradley, and M. Hughes (1998), Global-scale temperature patterns and climate forcing over the past six centuries, *Nature*, 392, 779–787.

Mann, M. E., R. Bradley, and M. Hughes (1999), Northern Hemisphere temperatures during the past millennium: inferences, uncertainties and limitations, *Geophys. Res. Lett.*, 26, 759–762.

Mann, M., E. S. Rutherford, E. Wahl, and C. Ammann (2005), Testing the fidelity of methods used in proxy-based reconstructions of past climate, *J. Climate*, 18, 4097–4107.

Mann, M. E., S. Rutherford, E. Wahl, and C. Ammann (2007), Reply to Comments on testing the fidelity of methods used in proxy-based reconstructions of past climate by Smerdon and Kaplan, *J. Climate*, 20, 5671–5674.

Mann, M. E., S. Rutherford, E. Wahl, and C. Ammann (2007), Robustness of proxy-based climate field reconstruction methods, *J. Geophys. Res.*, 112, D12109, doi:10.1029/2006JD008272, 2007.

Mann, M. E., Z. Zhang, M. Hughes, R. Bradley, S. Miller, and S. Rutherford (2008), Proxy-based reconstructions of hemispheric and global surface temperature variations over the past two millennia, *Proc. Natl. Acad. Sci.* 105, 13252–13257.

Mann, M. E., Z. Zhang, S. Rutherford, R. Bradley, M. Hughes, D. Shindell, C. Ammann, G. Faluvegi, and F. Ni (2009), Global signatures and dynamical origins of the Little Ice Age and Medieval Climate Anomaly, *Science*, 326, 1256–1260.

Mann, M., J. Fuentes, and S. Rutherford (2012), Underestimation of volcanic cooling in tree-ring-based reconstructions of hemispheric temperatures, *Nature Geoscience*, 5, 202–205.

Melvin, T. (2004), Historical growth rates and changing climatic sensitivity of boreal conifers, Ph.D. thesis, Climatic Research Unit, East Anglia, UK.

Melvin, T. M., and K. Briffa (2008), A "Signal-Free" approach to dendroclimatic Standardization, *Dendrochronologia*, 26, doi:10.1016/j.dendro.2007.12.001 71–86.

Melvin, T., H. Grudd, and K. Briffa (2012), Potential bias in "updating" tree-ring chronologies using regional curve standardisation: reprocessing 1500 years of Tornetrask density and ring-width data, *The Holocene*, doi:10.1177/0959683612460791.

Milakovsky, B., B. Frey, and T. James (2012), Carbon dynamics in the boreal forest, pp.109–135, In: Managing Forest Carbon in a Changing Climate, Eds.: M. Ashton, M. Tyrrell, D. Spalding, and B. Gentry, Springer.

Moberg, A., D. Sonechkin, K. Homgren, N. Datsenko, and W. Karlen (2005), Highly variable Northern Hemisphere temperatures reconstructed from low-and high-resolution proxy data, *Nature*, 433, 613–617, doi:10.1038/nature03265.

Nash, S. (2000), James Louis Giddings' archaeological tree-ring dating in the American Arctic: A forgotten legacy, *Arctic Anthropology*, 37, 60–78.

National Research Council, (2006), Report on Surface Temperature Reconstructions for the Last 2,000 Years, June 22, 2006. http://nationalacademies.org.

Naurzbaev, M. M., and E. A. Vaganov (1999), 1957-year tree-ring chronology of the eastern part of Taymir, *Sib. J. Ecol.*, 6, 159–168.

Nicolussi, K., and P. Schiessling (2001), Establishing a multi-millennial*Pinus cembra* chronology for the central Eastern Alps, in International Conference of Tree-Rings and People, edited by M. Kaennel-Dobbertin and O. U. Braker, pp. 251–252, *Swiss Fed. Res. Inst. WSL*, Davos, Switzerland.

Nishimura, P.H., and C. Laroque, (2010), Tree-ring evidence of larch sawfly outbreaks in western Labrador, *Canada, Canadian Journal of Forest Research*, 40, 1542–1549.

Oquilluk, W. A. (1973), People of Kauwerak: Legends of the Northern Eskimo. The third disaster (Ch. 5),With the assistance of Laurel L. Bland, Alaska Methodist University, Alaska Pacific University Press.

Osborn, T. J., and K. Briffa (2006), The spatial extent of 20th century warmth in the context of the last 1200 years, *Science*, 311, 841–844.

Pan, Y., R. A. Birdsey, J. Fang, R. Houghton, P. E. Kauppi, W. A. Kurz, O. L. Phillips, A. Shvidenko, S. L. Lewis, J. G. Canadell, P. Ciais, R. B. Jackson, S. Pacala, A. D. McGuire, S. Piao, A. Rautiainen, S. Sitch, and D. Hayes (2011), A large and persistent carbon sink in the world's forests, *Science*, doi:10.1126/science.1201609.

Panyushkina, I. P., D. V. Ovtchinnikov, and M. F. Adamenko (2005), Mixed response of decadal variability in larch tree-ring chronologies from upper tree-lines of Russian Altai, *Tree-Ring Research*, 61, (1), 33–42.

Payette, S. (2007), Contrasted dynamics of northern Labrador tree lines caused by climate change and migrational lag, *Ecology*, 88, 770–780.

Payette, S., and L. Filion (2010), (Editors), The Dendroecology: Principle, Methods and Application, University Of Laval Press, Quebec, Canada. 758 pages.

Pisaric, M. F. J., S. Carey, S. Kokelj, and D. Youngblut (2007), Anomalous 20th century tree growth, Mackenzie Delta, Northwest Territories, Canada, *Geophys. Res. Lett.*, 34, L05714, doi:10.1029/2006GL029139.

Porter, T., and M. Pisaric (2011), Temperature-growth divergence in white spruce forests of Old Crow Flats, Yukon Territory, and adjacent regions of northwestern North America, *Global Change Biology*, 11, 3418–3430.

Porter, T. J., M. Pisaric, S. Kokelj, and T. Edwards (2009), Climatic signals in δ13C and δ18O of tree-rings from white spruce in the Mackenzie Delta region, northern Canada, Arctic, *Antarctic and Alpine Research*, 41(4), 497–505.

Porter, T., M. Pisaric, S. Kokelj, and P. deMontigny (2013), A ring-width-based reconstruction of June-July minimum temperatures since AD 1245 from white spruce stands in the Mackenzie Delta region, northwestern Canada, Quaternary Research, doi:10.1016/j.yqres.2013.05.004.

Rutherford, S., M. Mann, T. Osborn, R. Bradley, K. Briffa, M. Hughes, and P. Jones (2005), Proxy-based Northern Hemisphere surface temperature reconstructions: sensitivity to method, predictor network, target season, and target domain, *J. Climate*, 18, 2308–2329.

Rydval, M., and R. Wilson (2012), The impact of industrial SO_2 pollution on north Bohemia conifers, *Water, Air, Soil and Pollution*, 223 (9), 5722–2744, doi:10.1007/s11270-012-1310-6.

Schulman, E. (1944), Tree-ring work in Scandinavia, *Tree-Ring Bulletin*, 11, 2–6.

Schweingruber, F. (1988), Tree Rings. Basics and Applications of Dendrochronology, Kluwer, Dordrecht. 276 pp.

Shen, C., W. Wang, W. Gong, and Z. Hao (2006), A Pacific Decadal Oscillation record since 1470 AD reconstructed from proxy data of summer rainfall over eastern China, *Geophys. Res. Lett.*, 33, L03702, February. doi:10.1029/2005GL024804.

Smerdon, J. E., and A. Kaplan (2007), Comments on "Testing the fidelity of methods used in proxy-based reconstructions of past climate": The role of the standardization interval, *J. Climate*, 20, 5666–5670.

Smerdon, J. E., A. Kaplan, D. Chang, and M. N. Evans (2010), A pseudoproxy evaluation of the CCA and RegEM Methods for reconstructing climate fields of the last millennium, *J. Climate*, 23, doi:10.1175/2010JCLI3328.1.

Solomina, O., and K. Alverson (2004), High latitude Eurasian paleo environments: Introduction and synthesis, *Palaeogeography, Palaeoclimatology, Palaeoecology*, 209, 1–18.

Solomina, O., G. Wiles, T. Shiraiwa, and R. D'Arrigo (2006), Multiproxy records of climate variability in Kamchatka over the last 400 years and initial comparisons with Alaska, *Climate of the Past*, Special Issue 2, 1051–1073.

Speer, J. (2012), Fundamentals of Tree-Ring Research. University of Arizona Press. 324 pp.

Stokes, M., and T. Smiley (1968), An Introduction to Tree-Ring Dating, University of Arizona Press, Tucson.

Thompson, D. W. J., J. Kennedy, J. Wallace, and P. Jones (2008), A large discontinuity in the mid-twentieth century in observed global-mean surface temperature, *Nature*, 453, 646–649.

Timonen, M. (2007), Past, Present and Future Climate from Tree Rings. Studies on past and present climate change in Finnish Lapland, based on the 7641-yr supra-long pine chronology for Finnish Lapland. Pdf (13 MB). Forest Science Day 23-Oct-2007, arranged by The Finnish Society of Forest Science and Taksaattoriklubi, themed as "Adaptation to climate change and Finnish Forests." *The House of Sciences (Tieteiden talo)*, Kirkkokatu 6, Helsinki, Finland.

Tingley, M., and P. Huybers (2010a), A Bayesian algorithm for reconstructing climate anomalies in space and time. Part 1: Development and applications to paleoclimate reconstruction problems, *J. Climate*, 23(10), 2759– 2781, doi:10.1175/2009JCLI3015.1.

Tingley, M., and P. Huybers (2010b), A Bayesian algorithm for reconstructing climate anomalies in space and time. Part 2: Comparison with the Regularized Expectation-Maximization Algorithm, *J. Climate*, 23(10), 2782–2800, doi:10.1175/2009JCLI3016.1.

Torrence, C., and G. Compo (1998), A Practical Guide to Wavelet Analysis, *Bull. Amer. Met. Soc.*, 79, 61–78.

Tranquillini, W. (1979), Physiological Ecology of the Alpine Timberline, Springer-Verlag, New York.

Trouet, V., J. Esper, A. Baker, and E. Zorita (2009), Synoptic-scale climate dynamics over the last millennium: a case study for the MCA-LIA transition, European Science Foundation Scientific Report. 17–20 May 2009.

US Global Change Research Program-Regional Climate Impacts—Alaska Report, (2009), pp. 139–144. Cambridge University Press, NY.

Vaganov, E., M. Hughes, A. Kirdyanov, F. Schweingruber, and P. Silkin (1999), Influence of snowfall and melt timing on tree growth in Subarctic Eurasia, *Nature*, 400, 149–151.

Vaganov, E. A., M.K. Hughes, and A.V. Shashkin (2006), Growth dynamics of tree rings: Images of past and future environments, Springer-Verlag, Berlin/Heidelberg/ New York.

Villalba, R., B. Luckman, J. Boninsegna, R. D'Arrigo, A. Lara, J. Villanueva-Diaz, M. Masiokas, J. Argollo, C. Soliz, C. LeQuesne, D. Stahle, F. Roig, J. Aravena, M. Hughes, G. Wiles, G. Jacoby, P. Hartsough, R. Wilson, E. Watson, E., E. Cook, J. Cerano-Paredes, M. Therrell, M. Cleaveland, M. Morales, M. Salzer, J. Moya, J. Pacajes, G. Massacchesi, F. Biondi, R. Urrutia, and G. Martinez Pastur (2011), Dendroclimatology from regional to continental scales: Understanding regional processes to reconstruct large-scale climatic variations across the Western Americas, In:Tree Rings and Climate: Sharpening the Focus, M. Hughes, H. Diaz, and T. Swetnam (editors), Advances in Global Change Research, Springer.

Visser, H., and J. Molenaar (1990), Detecting time-dependent climatic responses in tree rings using the Kalman filter, In: Cook, E., Kairiukstis, L., (Eds.), Methods of Dendrochronology, Kluwer Academic Press, Dordrecht, pp. 270–277.

von Storch, H., E. Zorita, J. Jones, Y. Dimitriev, F. Gonzalez-Rouco, and S. Tett (2004), Reconstructing past climate from noisy data, *Science*, 306, 679–682, doi:10.1126/ science.1096109.

von Storch, H., E. Zorita, and J. F. González-Rouco (2009), *International Journal of Earth Sciences* (Geol. Rundsch.), 98, 67–82. 10.1007/s00531-008-0349-5.

Wahl, E., and C. Ammann (2007), Robustness of the Mann, Bradley, Hughes reconstruction of Northern Hemisphere surface temperatures: examination of criticisms based on the nature and processing of proxy climate evidence, *Climatic Change*, 85, 33–69.

Wahl, E., D. Anderson, B. Bauer, R. Buckner, E. Gille, W. Gross, M. Hartman, and A. Shah (2010), An archive of high-resolution temperature reconstructions over the past 2+ millennia, *Geochemistry, Geophysics and Geosystems*, 11,14 January 2010. Q01001, doi:10.1029/2009GC002817.

Wang, L., S. Payette, and Y. Begin (2001), 1300-year tree-ring width and density series based on living, dead and subfossil black spruce at treeline in Subarctic Quebec, *The Holocene*, 11, 333–341.

Wiles, G., D. Barclay, P. Calkin, and T. Lowell (2008), Century to millennial-scale temperature variations for the last two thousand years indicated from glacial geologic records of southern Alaska, *Global and Planetary Change*, 60, 115–125.

Wiles, G. C., D. Lawson, E. Lyon, and N. Wiesenberg (2011), Tree-ring dates on two pre-Little Ice Age advances in Glacier Bay National Park and Preserve, Alaska, USA, *Quaternary Research*, doi:10.1016/j.yqres.2011.05.005.

Wilmking, M., R. D'Arrigo, G. Jacoby, and G. Juday (2005), Divergent growth responses in circumpolar boreal forests, *Geophys. Res. Lett.*, 32, L15715, 10.1029/2005GLO23331.

Wilson, R., and B. Luckman (2002), Tree-ring reconstruction of maximum and minimum temperatures and the diurnal temperature range in British Columbia, Canada, *Dendrochronologia*, 20, 1–12.

Wilson, R. J. S., and B. Luckman (2003), Dendroclimatic reconstruction of maximum summer temperatures from upper tree-line sites in interior British Columbia, *The Holocene*, 13, 853–863.

Wilson, R., and W. Elling (2004), Temporal instability in tree-growth/climate response in the Lower Bavarian Forest region: implications for dendroclimatic reconstruction, *Trees*, 18, 19–28.

Wilson, R. J. S., and J. Topham (2004), Violins and climate, *Theor. Appl. Climatol.*, 77, 9–24.

Wilson, R., N. Loader, M. Rydval, H. Paton, A. Frith, C. Mills, A. Crone, C. Edwards, L. Larsson, and B. Gunnarson (2011), Reconstructing Holocene climate from tree rings—the potential for a long chronology from the Scottish Highlands, *The Holocene*, doi:10.1177/0959683611405237.

Xu, L., R. Myneni, F. Chapin III, T. Callaghan, J. Pinzon, C. Tucker, Z. Zhu, J. Bi, P. Ciais, H. Tommervik, E. Euskirchen, B. Forbes, S. Piao, B. Anderson, S. Ganguly, R. Nemani, S. Goetz, P. Beck, A. Bunn, C. Cao, and J. Stroeve (2013), Temperature and vegetation seasonality diminishment over northern lands, *Nature Clim. Change*, 3, 581–586, doi:510.1038/nclimate1836.

Youngblut, D., and B. Luckman (2008), Maximum June–July temperatures in the Southwest Yukon over the last 300 years reconstructed from tree rings, *Dendrochronologia*, 25, 153–166.

CORE TRL-LDEO PUBLICATIONS ON NORTHERN FORESTS

The following references are authored or co-authored by members of the TRL-LDEO and collaborators, and are broadly relevant to tree-ring based Northern Hemisphere temperature and climatic reconstructions and related issues.

2010–Present

Anchukaitis, K., K. Briffa, A. Buchwal, U. Büntgen, E. Cook, R. D'Arrigo, J. Esper, M. Evans, D. Frank, H. Grudd, B. Gunnarson, M. Hughes, A. Kirdyanov, P. Krusic, B. Luckman, T. Melvin, M. Salzer, C. Timmreck, E. Vaganov, and R. Wilson (2012a), Tree rings and volcanic cooling, *Nature Geoscience*, 5, 836–837.

Anchukaitis, K., R. D'Arrigo, L. Hayles, D. Frank, A. Verstege, B. Buckley, A. Curtis, G. Jacoby, and E. Cook, (2012b), A non-diverging tree-ring density reconstruction of summer temperatures from northwestern North America, *Journal of Climate*, doi:10.1175/JCLI-D-11-00139.1, 2012.

Andreu-Hayles, L., R. D'Arrigo, K. Anchukaitis, P. Beck, D. Frank, A. Verstege, and S. Goetz (2011), Varying northern forest response to Arctic environmental change at the Firth River, Alaska, Special issue on Arctic Vegetation Dynamics, *Environ. Res. Lett.*, 6, 045503 doi:10.1088/1748-9326/6/4/045503.

D'Arrigo, R., R. Seager, J. Smerdon, A. LeGrande, and E. Cook (2011), The anomalous winter of 1783–4: was the Laki eruption or an analog of the 2009/10 winter to blame? *Geophys. Res. Lett.*, 38, L05706, doi:10.1029/2011GL046696.

D'Arrigo, R., K. Anchukaitis, B. Buckley, E. Cook, and R. Wilson (2011), North Atlantic Oscillation climate signature in West Virginia red cedar over the past millennium, *Global and Planetary Change*, 84, 8–13, Special Issue on Perspectives on Climate in Medieval Time, doi:10.1016/j.gloplacha.2011.07.003.

D'Arrigo, R., R. Wilson, and K. Anchukaitis (2013), Volcanic cooling signals in tree rings for the past millennium, In press, *J. Geophys. Res. Atmospheres*.

Davi, N., G. Jacoby, K. Fang, J. Li, R. D'Arrigo, N. Baatarbileg, and D. Robinson (2010), Reconstructed drought across Mongolia based on a large-scale tree-ring network: 1520–1993, *J. Geophys. Research*, 15, doi:10.1029/2010JD013907.

Dendroclimatic Studies: Tree Growth and Climate Change in Northern Forests,
Monograph Number 67. Rosanne D'Arrigo, Nicole Davi, Gordon Jacoby,
Rob Wilson, and Greg Wiles.

Visser, H., U. Büntgen, R. D'Arrigo, and A. Petersen (2010), Detecting instabilities in tree-ring proxy calibration, *Climate of the Past Discussions*, 6, 225–255, doi:10.5194/cpd-6-225-2010.

Wiles, G. C., D. Lawson, E. Lyon, and N. Wiesenberg (2011), Tree-ring dates on two pre-Little Ice Age advances in Glacier Bay National Park and Preserve, Alaska, USA, *Quaternary Research*, doi:10.1016/j.yqres.2011.05.005.

Wiles, G. C., C. Mennett, S. Jarvis, R. D'Arrigo, N. Wiesenberg, and D. Lawson (2012), Tree-ring investigations into changing climatic responses of Yellow-Cedar, Glacier Bay, Alaska, *Can. J. Forest Research*, DOI: 10.1139/X2012-028.

Wilson, R., N. Loader, M. Rydval, H. Paton, A. Frith, C. Mills, A. Crone, C. Edwards, L. Larsson, and B. Gunnarson (2011), Reconstructing Holocene climate from tree rings—the potential for a long chronology from the Scottish Highlands, *The Holocene*, 22, 3 11. DOI: 10.1177/0959683611405237.

2009

D'Arrigo, R., G. Jacoby, B. Buckley, J. Sakulich, D. Frank, R. Wilson, A. Curtis, and K. Anchukaitis (2009), Tree growth and inferred temperature variability at the North American Arctic treeline, *Global and Planetary Change*, 65, 71–82, doi:10.1016/j.gloplacha.2008.10.011.

2008

D'Arrigo, R., R. Wilson, B. Liepert, and P. Cherubini (2008), On the "divergence problem" in northern forests: a review of the tree-ring evidence and possible causes, *Global and Planetary Change*, 60, 289–305; doi:10.1016/j.gloplacha.2007.03.004.

2007

Wilson, R., G. Wiles, R. D'Arrigo, and C. Zweck (2007a), Cycles and shifts: 1300-years of multi-decadal temperature variability in the Gulf of Alaska, *Climate Dynamics*, 28, 425–440.

Wilson, R., R. D'Arrigo, B. Buckley, U. Büntgen, J. Esper, D. Frank, B. Luckman, S. Payette, S. Vose, and D. Youngblut (2007b), A matter of divergence—tracking recent warming at hemispheric scales using tree-ring data, JGR-Atmospheres, 112, D17103, doi:10.1029/2006JD008318.

2006

D'Arrigo, R., R. Wilson, and G. Jacoby (2006), On the long-term context for late twentieth century warming, *J. Geophys. Res.*, 111, D03103, doi:10.1029/2005JD006352.

2005

D'Arrigo, R., E. Mashig, D. Frank, R. Wilson, and G. Jacoby (2005), Temperature variability over the past millennium inferred from northwestern Alaska tree rings, *Climate Dynamics*, 10.1007/s00382-004-0502-1.

D'Arrigo, R., G. Jacoby, R. Wilson, and F. Panagiotopoulos (2005), A reconstruction of the Siberian High since AD1599 from Eurasian and North American tree rings, *Geophys. Res. Lett.*, 32, L05705, 10.1029/2004GL022271.

D'Arrigo, R., R. Wilson, C. Deser, G. Wiles, E. Cook, R. Villalba, S. Tudhope, J. Cole, and B. Linsley (2005), Tropical-North Pacific climate linkages over the past four centuries, *Journal of Climate*, 18, 5253–5265, Dec 15, 2005.

Driscoll, W., G. Wiles, R. D'Arrigo, and M. Wilmking (2005), Dendroclimatic evidence of changing environmental stresses on tree growth, Lake Clark National Park and Preserve, Alaska, *Geophys. Res. Lett.*, 32, L20703; 10.1029/2005GL024258.

2004

Cook, E., J. Esper, and R. D'Arrigo (2004), Extra-tropical Northern Hemisphere temperature variability over the past 1000 years, *Quat. Sci. Rev.*, 23, 2063–2074.

D'Arrigo, R., E. Mashig, D. Frank, G. Jacoby, and R. Wilson (2004), Reconstructed warm season temperatures for Nome, Seward Peninsula, Alaska since AD 1389, *Geophys. Res. Lett.*, 31, L09202, doi:10.1029/2004GL019756.

D'Arrigo, R., R. Kaufmann, N. Davi, G. Jacoby, C. Laskowski, and R. Myneni (2004), Thresholds for warming-induced growth decline at elevational treeline in Yukon Territory, Canada, *Global Biogeochem. Cycles*, 18, doi:10.1029/2004GBO02249.

Jacoby, G., O. Solomina, D. Frank, N. Eremenko, and R. D'Arrigo (2004), Kunashir (Kurile Islands) oak response to temperature and relation to the Pacific Decadal Oscillation (PDO), *Palaeogeography, Palaeoclimatology, Palaeoecology*, 209, 303–311.

Wiles, G., R. D'Arrigo, R. Villalba, and P. Calkin (2004), Solar forcing of century-scale mountain glaciations over the past millennium in Alaska, *Geophys. Res. Lett.*, 31, L15203, doi:10.1029/2004GL020050.

2003

D'Arrigo, R., E. Cook, M. E. Mann, and G. Jacoby (2003), Tree-ring reconstruction of temperature and sea-level variability associated with the warm season Arctic Oscillation since AD 1650, *Geophy. Res. Lett.*, 30, 1549.

D'Arrigo, R., B. Buckley, and S. Kaplan (2003), Interannual to multidecadal modes of variability in northern Labrador tree rings spanning the past four centuries, *Climate Dynamics*, 20, 219–228.

D'Arrigo, R., G. Jacoby, and D. Frank (2003), Dendroclimatological evidence for major volcanic events of the past two millennia. pp. 255–263 in AGU Geophysical Monograph 139, Chapman Conference on Volcanism, Santorini. A. Robock and C. Oppenheimer, Editors.

Davi, N., G. Jacoby, and G. Wiles (2003), Boreal temperature variability inferred from maximum latewood density and tree-ring width data, Wrangell Mountain region, Alaska, *Quat. Res.*, 60, 252–262.

Jacoby, G., and R. D'Arrigo (2003), The Laki eruption and observed dendroclimatic effects of volcanism. pp. 263–272 in Chapman Conference on Volcanism and the Earth's Atmosphere, Santorini, Greece, June 17–21 2002 and proceedings volume for *Chapman Monograph*.

Jacoby, G., N. Pederson, and R. D'Arrigo (2003), Temperature and precipitation in Mongolia based on dendroclimatic investigations, *Chinese Sci. Bull.*, 48, 1474–79.

2002

Cook, E., R. D'Arrigo, and M. Mann (2002), A well-verified, multi-proxy reconstruction of the winter North Atlantic Oscillation index since AD 1400, *J. of Climate*, 15, 1754–1764.

Davi, N., R. D'Arrigo, G. Jacoby, B. Buckley, and O. Kobayashi (2002), Warm-season annual temperature variability for Hokkaido, Japan reconstructed from tree-ring density data: AD 1557–1990, *Climatic Change*, 52, 201–217.

2001

D'Arrigo, R., R. Villalba, and G. Wiles (2001), Tree-ring estimates of Pacific decadal climate variability, *Climate Dynamics*, 18, 219–224.

D'Arrigo, R., G. Jacoby, D. Frank, N. Pederson, E. Cook, B. Buckley, Baatarbileg Nachin, R. Mijiddorj, and C. Dugarjav (2001a), 1738 years of Mongolian temperature variability inferred from a tree-ring record of Siberian pine, *Geophys. Res. Lett.*, 28, 543–546.

D'Arrigo, R., D. Frank, G. Jacoby, and N. Pederson (2001b), Spatial response to major volcanic events on or about AD 536, 934 and 1258: Frost rings and other dendrochronological evidence from Mongolia, *Climatic Change*, 49, 239–46.

Pederson, N., G. Jacoby, R. D'Arrigo, and B. Buckley (2001), Hydrometeorological reconstructions for northeastern Mongolia derived from tree rings: 1651–1995, *J. of Climate*, 14, 872–881.

2000

D'Arrigo, R. D., C. M. Malmstrom, G. C. Jacoby, S. Los, and D. E. Bunker (2000), Tree-ring indices of interannual biospheric variability: testing satellite-based model estimates of vegetation activity, *Internat. J. Remote Sens.*, 21, 2329–2336.

D'Arrigo, R., G. Jacoby, N. Pederson· D. Frank, B. Buckley, N. Baatarbileg , R. Mijiddorj, and C. Dugarjav (2000), Mongolian tree rings, temperature sensitivity and reconstructions of Northern Hemisphere temperature, *The Holocene*, 10, 669–672.

Jacoby, G., N. Lovelius, O. Shumilov, O. Raspopov, J. Kurbainov, and D. Frank (2000), Long-term temperature trends and tree growth in the Taymir region of northern Siberia, *Quat. Res.*, 53, 312–318.

1999

D'Arrigo, R. (1999), On the Thelon River, In: Lamont-Doherty Earth Observatory: Twelve Perspectives on the First Fifty Years (1949–1999), L. Lippsett, Ed., Columbia Univ. Press (New York).

D'Arrigo, R. D., and G. C. Jacoby (1999), Northern North American tree-ring evidence for regional temperature changes after major volcanic events, *Climatic Change*, 41, 1–15.

Jacoby, G. (1999), Using trees to reveal Earth's environmental history. In: Lamont-Doherty Earth Observatory: Twelve Perspectives on the First Fifty Years (1949–1999), L. Lippsett, Ed., Columbia Univ. Press (New York).

Jacoby, G. C., K. Workman, and R. D'Arrigo (1999), 1783 Laki eruption, tree rings and catastrophe for northwestern Inuit, *Quat. Sci. Rev.*, 18, 53–59.

Jacoby, G. C., R. D'Arrigo, and G. Juday (1999), Tree-ring indicators of climate change at northern latitudes, *World Resources Review*, 11, 21–29.

D'Arrigo, R. D., G. Jacoby, M. Free, and A. Robock (1999), Northern Hemisphere temperature variability for the past three centuries: tree-ring and model estimates, *Climatic Change*, 42, 663–675.

D'Arrigo, R., G. Wiles, G. Jacoby, and R. Villalba (1999), North Pacific sea surface temperatures: past variations inferred from tree rings, *Geophysical Research Lett.*, 26, 2757–2760.

1998

Wiles, G. C., R. D. D'Arrigo, and G. C. Jacoby (1998), Gulf of Alaska atmosphere-ocean variability over recent centuries inferred from coastal tree-ring records, *Climatic Change*, 38, 289–306.

1997

Jacoby, G., and R. D'Arrigo (1997), Tree rings, carbon dioxide and climatic change. *Proc. Nat. Acad. Sci.*, 94, 8350–8353.

D'Arrigo, R. D., D. K. Yamaguchi, G. C. Wiles, G. C. Jacoby, A. Osawa, and D. M. Lawrence (1997), A kashiwa oak tree-ring width chronology from northern coastal Hokkaido, Japan, *Can. J. For. Res.*, 27, 613–617.

1996

D'Arrigo, R. D., E. R. Cook, and G. C. Jacoby (1996), Annual to decadal-scale variations in northwest Atlantic sector temperatures inferred from Labrador tree rings, *Canadian J. For. Res.*, 26, 143–148.

Gostev, M., G. Wiles, R. D'Arrigo, and G. Jacoby (1996), Temperature reconstruction for Kamchatka based on Siberian larch tree-ring data, *Canadian Journal of Forest Research*, 26, 2048–2052.

Jacoby, G. C., R. D. D'Arrigo, and Ts. Davaajamts (1996), Mongolian tree rings and twentieth century warming, *Science*, 273, 771–773.

Jacoby, G. C., G. C. Wiles, and R. D. D'Arrigo (1996), Alaskan dendroclimatic variations for the past 300 years along a north-south transect, *Radiocarbon*, pp. 235–248.

D'Arrigo, R. D., and G. C. Jacoby (1996), Past Changes in Climate and Tree Growth in the Western Americas. In: Lawford, G., P. B. Alaback and E. Fuentes, Eds. High-Latitude Rainforests and Associated Ecosystems of the West Coast of the Americas. Climate, Hydrology, Ecology and Conservation. Springer-Verlag, New York. AMIGO Conference, Sept. 21–26, Vancouver Island, pp. 75–88.

Jacoby, G. C., R. D. D'Arrigo, and B. H. Luckman (1996), Millenial and near-millennial dendroclimatic studies in northern North America, Proceedings of the NATO Advanced Research Workshop, Climate Variations and Forcing Mechanisms of the Last 2,000 Years. Il Ciocco, Lucca, Italy, October 3–7, 1994, In: Jones, P. D., Bradley, R. S., and J. Jouzel, Eds. *NATO ISI Series* I41: 67–84.

Wiles, G. C., R. D. D'Arrigo, and G. C. Jacoby (1996), Summer temperature changes along the Gulf of Alaska and the Pacific Northwestern coast modeled from coastal tree rings, *Canadian Journal of Forest Research*, 26, 474–481.

1995

Jacoby, G. C., and R. D. D'Arrigo (1995), Tree-ring width and density evidence of climatic and potential forest change in Alaska, *Global Biogeochemical Cycles*, 9, 227–234 and related *Science* article: G. Taubes (1995), Is a warmer climate wilting the forests of the north? *Science*, 267, 1595.

1993

D'Arrigo, R. D., and G. C. Jacoby (1993), Tree growth-climate relationships at the northern boreal forest treeline of North America: evaluation of potential response to increasing carbon dioxide, *Global Biogeochemical Cycles*, 7, 525–536.

D'Arrigo, R. D., E. R. Cook, G. C. Jacoby, and K. R. Briffa (1993), NAO and sea surface temperature signatures in tree-ring records from the North Atlantic sector, *Quaternary Science Reviews*, 12, 431–440.

D'Arrigo, R. D., and G. C. Jacoby (1993), Secular trends in high northern-latitude temperature reconstructions based on tree rings, *Climatic Change*, 25, 163–177.

1992

D'Arrigo, R. D., G. C. Jacoby, and R. M. Free (1992), Tree-ring width and maximum latewood density at the North American treeline: parameters of climatic change, *Canadian Journal of Forest Research*, 22, 1290–1296.

D'Arrigo, R. D., and G. Jacoby (1992), Dendroclimatic evidence from northern North America. In: "Climate since A.D. 1500," R. S. Bradley and P. D. Jones, Eds., pp. 296–311. Routledge, London.

D'Arrigo, R. D., G. C. Jacoby, and E. R. Cook (1992), Impact of recent North Atlantic anomalies on surrounding land areas based on dendroclimatic evidence, *Geophysical Research Letters*, 19, 2321–2324.

Jacoby, G. C., and R. D'Arrigo (1992), Global change and thermal history as recorded by northern North American tree-ring data, Proceedings of International Conference on the Role of Polar Regions in Global Change, Univ. of Alaska, Fairbanks. pp. 599–602.

Jacoby, G. C., and R. D'Arrigo (1992), Spatial patterns of tree growth anomalies from the North American boreal treeline in the early 1800's, including the year 1816, in The Year without a Summer?: World Climate in 1816, Harington, C.R., Ed. Natl. Museum of Science, Ottawa, Canada, 225–265.

1991

Thetford, R. D, R. D. D'Arrigo, and G. C. Jacoby (1991), An image analysis system for generating densitometric and ring width time series, *Canadian Journal of Forest Research*, 21, 1544–1549.

1989

Jacoby, G. C., and R. D'Arrigo (1989), Reconstructed Northern Hemisphere annual temperature since 1671 based on high latitude tree-ring data from North America, *Climatic Change*, 14, 39–59.

1988

Jacoby, G., I. Ivanciu, and L. Ulan (1988), A 263-year record of summer temperature for northern Quebec reconstructed from tree-ring data and evidence of a major climatic shift in the early 1800s, *Palaeogeography, Palaeoclimatology, Palaeoecology*, 64, 69–78.

1987

D'Arrigo, R. D., Jacoby, G. C., and I. Y. Fung (1987), Boreal forests and atmosphere-biosphere exchange of carbon dioxide, *Nature*, 329, 321–323.

1985

Jacoby, G., E. Cook, and L. Ulan (1983), Reconstructed summer degree days in central Alaska and Northwestern Canada since 1524, *Quat. Res.*, 23, 18–26.

1982

Jacoby, G., and L. Ulan (1982), Reconstruction of past ice conditions in a Hudson Bay Estuary using tree rings, *Nature*, 298, 637–639.

1981

Jacoby, J., and E. Cook (1981), Past temperature variations inferred from a 400-year tree-ring chronology from Yukon Territory, Canada. *Arctic and Alpine Res.*, 13, 409–418.

Jacoby, G., and L. Ulan (1981), Review of dendroclimatology in the forest-tundra ecotone of Alaska and Canada. In: Syllogeus: Climatic Change in Canada—2, C. R. Harington, Ed., V. 33, 97–128. *Nat. Mus. Canada*, Ottawa.

Related Reports

TRL-LDEO tree-ring results featured in the following reports.

Jacoby, G. C., and R. D. D'Arrigo (1992), Indicators of Climatic and Biospheric Change: Evidence from Tree-Rings. In: Senate Hearing on Global Change Research: Indicators of Global Warming and Solar Variability, before the Commission on Commerce, Science and Transportation, U.S. Senate, 122nd Congress, Second Session. U.S. Government Printing Office. Feb. 27, 1992. Senator Al Gore, presiding.

D'Arrigo, R. (2007), Presenter, Results included in National Research Council Report On Surface Temperature Reconstructions of the past 1,000–2,000 Years.

Jansen, E., J. Overpeck, K. R. Briffa, J.–C. Duplessy, F. Joos, V. Masson-Delmotte, D. Olago, B. Otto-Bliesner, W. R. Peltier, S. Rahmstorf, R. Ramesh, D. Raynaud, D. Rind, O. Solomina, R. Villalba, and D. Zhang. (2007), Intergovernmental Panel on Climatic Change (IPCC). The Physical Science Basis. Summary for Policy Makers.

INDEX

Page numbers followed by "f" and "t" indicate figures and tables.

Dendroclimatic Studies: Tree Growth and Climate Change in Northern Forests,
Monograph Number 67. Rosanne D'Arrigo, Nicole Davi, Gordon Jacoby,
Rob Wilson, and Greg Wiles.